PERFORMANCE CONTRACTING FOR ENERGY AND ENVIRONMENTAL SYSTEMS

PERFORMANCE CONTRACTING FOR ENERGY AND ENVIRONMENTAL SYSTEMS

Shirley J. Hansen, Ph.D.

Published by
THE FAIRMONT PRESS, INC.
700 Indian Trail
Lilburn, GA 30247

Library of Congress Cataloging-in-Publication Data

Hansen, Shirley J., 1928 -
Performance contracting for energy and environmental systems / Shirley J. Hansen.
p. cm.
Includes bibliographical references and index.
ISBN 0-88173-127-7
1. Heating and ventilating industry--United States. 2. Buildings--United States--Energy conservation. 3. Buildings--United States--Environmental engineering. 4. Buildings--United States--Performance. 5. Subcontracting--United States. I. Title.

HD9683.U52H36 1992 658.2'5-dc20 92-24389
CIP

Performance Contracting for Energy and Environmental Systems.

Published by The Fairmont Press, Inc.
700 Indian Trail
Lilburn, GA 30247

Printed in the United States of America

10 9 8 7 6 5 4 3 2 1

ISBN 0-88173-127-7 FP

ISBN 0-13-658915-4 PH

Distributed by PTR Prentice-Hall, Inc.
A Simon & Schuster Company
Englewood Cliffs, NJ 07632

Prentice-Hall International (UK) Limited, London
Prentice-Hall of Australia Pty. Limited, Sydney
Prentice-Hall Canada Inc., Toronto
Prentice-Hall Hispanoamericana, S.A., Mexico
Prentice-Hall of India Private Limited, New Delhi
Prentice-Hall of Japan, Inc., Tokyo
Simon & Schuster Asia Pte. Ltd., Singapore
Editora Prentice-Hall do Brasil, Ltda., Rio de Janeiro

CONTENTS

FOREWORD

Performance contracting erupted on the scene in the late 1970s and has since proved itself as a way for an organization to take advantage of energy cost saving opportunities. It is now a viable energy financing and service mechanism for any type of organization, for it *guarantees* results. References to "performance contracting" now appear in laws passed by the United States Congress and in many states. It has penetrated to the local level. More than 3,000 school districts across the nation have now entered into some type of performance contract.

This book is designed to provide you, and others in your organization, the guidance to use the performance contracting energy financing/ service mechanism effectively. It is directed at owners, business managers, facility managers and operators, who would like to provide a comfortable, productive environment *as cost-effectively as possible.* Part I builds an understanding of the options available and then walks you through each step to effectively secure performance contracting services.

Part II applies the material from Part I to three very distinctive market segments; i.e., the federal government, utilities and institutions. Performance contracting in each segment is examined from the end-users' and the energy service companies' points of view. Each part and segment stands alone, so you can pick and choose those parts that will benefit you the most.

Unless you are running a *very* efficient and cost-effective operation, this book is for you. For a commercial enterprise, performance contracting can provide the competitive edge. For public institutions, strapped for dollars, it frees money to meet critical needs. For utilities, it can help manage the demand-side of their operations. For the university, who's boiler won't carry it through another winter, here is the alternative to closing the doors. For those who have a crushing deferred maintenance burden, this is an opportunity to use energy cost savings to turn it around.

Throughout the book, insights are offered that may benefit energy service company managers and sales personnel. For ESCO personnel, however, the greatest benefit is apt to come from viewing the process from the customer's perspective.

Every situation is unique and the most successful programs will be tailored to *your* needs. The procedures discussed herein offer detailed guidelines for achieving tailor-made agreements.

Reference to any specific commercial product, process, or service by trade name, trademark, manufacturer, or otherwise, does not necessarily constitute or imply its endorsement, recommendation or favoring by the author.

The performance contracting industry is dynamic. Sorting through a range of options to bring you effective procedures and practices in use today was a most challenging task made easier by many people. In particular, I would like to acknowledge the very special insights, support and assistance from a few along the way, especially Ms. Patricia Rose, U.S. Department of Energy (DOE), who gave guidance to our DOE-funded development of the first national guidelines for "alternative financing;" Ms. Jeannie Weisman, who gave her unstinting support against the odds; and Mr. Robert J. Heller, who offered special insights on energy financing and encouraged the development of this book. For special guidance through the federal maze, my thanks go to Mr. Martin E. Nelson, San Diego Division, U.S. Postal Service and Mr. Doug Dahle, NAVFAC Shared Energy Savings Program Manager, U.S. Navy.

Providing consultation to clients also offers unparalleled opportunities to learn from them. My thanks to the many institutions, cities, states, firms and ESCOs who have shared with us the excitement of this growing industry. In particular, I wish to extend special appreciation to the Pennsylvania Energy Office for its support in creating an award winning manual, *The Bottom Line.* Much of the material presented in Part I has its roots in that manual.

My sincere appreciation goes to Ms. Terry Singer, Executive Director of the National Association of Energy Service Companies, for the leadership she has given performance contracting and for the time and thought she gave to critiquing portions of this book. Very special thanks and acknowledgements go to the contributing authors of Part II; Mr. J. Terry Radigan, Mr. Thomas K. Dreessen, Mr. David R. Wolcott, Mr. Cary G. Bullock, Mr. Joseph J. Lavorgna, Mr. James A. Smoyer and Mr. R. Scott Holland. Thanks, gentlemen, for sharing with us some exceptional insights that go with using performance contracting in some very unique markets. Finally, most heartfelt thanks to Mr. F. William Payne, Ms. Hope Worley and Mr. James C. Hansen, who made it possible.

Shirley J. Hansen

CHAPTER 1

THE ENERGY OPPORTUNITY

Energy efficiency is an investment; not an expense. Investing in energy efficiency can free dollars that are now being spent for wasted energy. Energy efficiency, in effect, generates "revenues" that can be redirected to new band uniforms, leasing an MRI, achieving a more competitive edge... or giving yourself a raise.

Once you start thinking of energy efficiency as an investment, you'll discover it's an investment that typically pays you back very handsomely. To use energy jargon: most retrofits have paybacks of four years or less. Translated to business vernacular that's a return on investment (ROI) of 25+ percent! It's hard to get a deal like that from the local bank.

It's similar to an income tax refund from the Internal Revenue Service. It's not really new money; it was already yours. However, it was committed to paying taxes (or utility bills), and now it's available to fulfill special wants.

There is, of course, an old truism that fits this case: "It takes money to make money." Setting aside dollars for energy work is not easy. Funds that could have been used for energy work are often diverted to needs more closely identified with the organization's mission. Math books, new carpeting in the lobby or bar code scanners frequently take precedence. As energy efficiency work is deferred, inefficiencies increase and the dollars going to the utility for wasted energy mount.

But there is hope. While you may not have money to invest in a more energy efficient operation, others do. They know you have $120,000 "hidden in your boiler," as indicated in the advertisement. Private sector firms have become aware of the millions languishing in America's "boiler rooms." A whole industry designed to serve this need now exists.

The process is simple. After inspecting a building for energy saving opportunities, an Energy Service Company (ESCO) will review the recommended energy conservation opportunities with the owner and implement those measures acceptable to the owner at no front end cost to the owner. The ESCO then *guarantees* that the energy savings will cover the cost of the capital modifications.

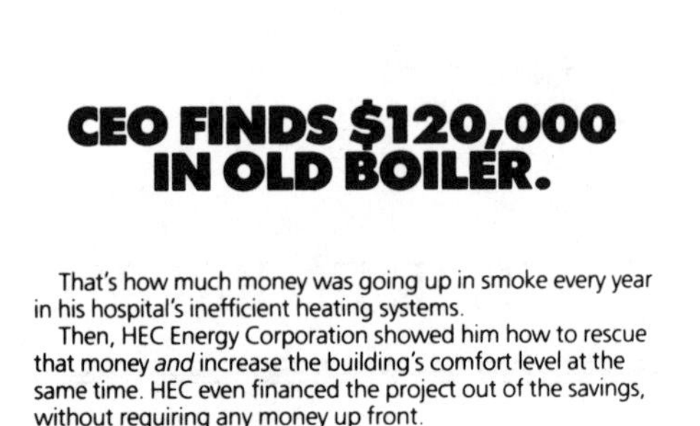

Performance Contracting in Retrospect

In its embryonic stages, in the late 1970s and early 1980s, the agreements for energy financing services generally turned on each party getting a percentage of the savings. The ESCO received a share to cover its costs and make a profit. The owner also received a share (as well as capital improvements) as an inducement to participate. Since each party received a share of the energy cost savings, this procurement procedure became known as "shared savings."

During the life of the contract, the ESCO expected its percentage of the cost savings to cover all the costs it had incurred, plus a profit. This concept worked quite well as long as energy prices stayed the same or escalated.

In the 1980s, energy prices began to drop. With lower prices, it took longer than predicted for the firm to recover its costs. Some firms could not meet their payments to their suppliers or financial backers. Companies closed their doors; and in the process, defaulted on their commitments to their shared savings partners. Some suppliers tried to recover costs from the building owners. Lawsuits were filed and "shared savings" was in trouble.

To make matters worse, it was discovered that one of the pioneers in the field had been entering into shared savings agreements with an eye toward benefiting primarily from federal investment tax credits and energy tax credits. The building owner, as the shared savings partner, did not necessarily receive any energy cost savings.

Once these stories started traveling, shared savings troubles just got bigger. The concept was sound, but the trust, so essential to a contract based on uncertain future savings, was badly shaken.

Fortunately, some agreements continued to show savings benefits to both parties. Of even greater importance, some companies had guaranteed the savings and they made good on those guarantees.

From this tenuous thread, the "shared savings" industry survived, but its character changed dramatically. Those supplying the financial backing and/or the equipment recognized the risks of basing contracts on energy price; interest rates went up. Insurance that had been available to ESCOs dwindled. By 1983, the true shared savings agreement had shrunk to approximately 5 percent of the market.

In its place, new names, new terms and new types of agreements emerged. New descriptive names for the process included alternative financing, positive cash flow financing, savings-based arrangements and performance contracting. Perhaps to respond to earlier "shared savings" fears, the industry focus turned to *guaranteed performance.* And performance contracting emerged as the favored name.

Different financing mechanisms also emerged, but nearly all of them offered savings guarantees of some type. Shared savings as a financing mechanism still existed, but its appeal was limited. Typically, projected savings were guaranteed to cover any debt service obligations that were incurred. Guaranteed savings through municipal lease, standard lease or installment purchase became the favored vehicles for financing energy efficiency projects.

Performance Contracting Today

The performance contracting industry has matured. Those who remain have track records that clearly demonstrate they can reliably meet the energy and financing needs of their customers. Unlike many other industries, energy service companies (ESCOs) really put their money where their mouths are. If you go to the doctor and he says I'm sorry but I don't know what the problem is, you still pay for his time—and his lack of knowledge. When an ESCO guarantees savings, it does not get paid unless those savings are achieved; i.e., unless the ESCO performs as predicted. That is the essence of *performance* contracting. The deck is stacked in the customer's favor. Typically, the only cost to the customer is money that would otherwise have gone to the utility for wasted energy.

As we move toward the 21st century, performance contracting is increasingly viewed as a viable way to use future energy savings to improve or replace equipment or modify the building envelope. Many view it as a way to operate a facility more cost-effectively. To some, it is viewed more narrowly as a way to "buy" needed equipment. A few see it as a way to free dollars to meet other needs. Performance contracting can be "all of the above."

The Changing "Package"

Initially, the attraction of shared savings and later, performance contracting, was financing. At that stage, all aspects focused on protecting the financial commitment. While financing is still a major attraction for prospective customers, many ESCOs have "unbundled" the package. They offer much more than financing. ESCOs provide expertise, equipment performance guarantees, maintenance, training, monitoring, savings verification and organizational momentum.

Some organizations, especially non-profits, have found they can secure their own financing at a lower interest rate. Under such an arrangement, the ESCO still serves a vital role as it can enhance the loan's terms by guaranteeing the savings.

The unbundled package offers a range of options from which you can select just what your organization needs. The take-it-or-leave-it ESCO contract is, fortunately, disappearing. ESCOs now realize that tailored programs make for happy partnerships—and successful projects.

If your organization, or division, has been frustrated by seeing the energy savings you labored to achieve going back to a general fund, a county budget or even to the state capital, performance contracting can help. A contract can be written to earmark some or all of your savings to meet specific needs; e.g., maintenance. ESCOs prefer to give you such an option, because you will have an incentive to fulfill your partnership commitment.

Probably the most crucial service the ESCO provides resides in its expertise. From years of experience, careful monitoring and verification procedures, ESCOs have learned exactly what works. Given certain conditions, they can predict with surprising accuracy, the level of savings that you can achieve by implementing specific measures.

Growing Acceptance

After a rocky beginning, there is little doubt that performance contracting works. It is especially attractive to non-profit institutions as they suffer from increasingly constrained budgets. They also have the added advantage of using tax-exempt status to secure more attractive terms.

At the federal level. In reauthorizing the federal energy grants program for schools and hospitals (P.L. 101-440), Congress specifically provided that performance contracting could be used for local matching funds. (See Chapter 20.)

In the waning days of pure "shared savings," Congress enabled federal agencies to use private sector financing through "shared energy savings" (SES). Unfortunately, statutes can freeze opportunities and the federal government is still locked into the "shared savings" approach. While SES offers some financial support and risk shedding for the government, it does not serve the federal government's best interests as well as other financing mechanisms might. (See Chapter 9, "Energy Financing Options," and Part II, Section 1, and Chapters 14 and 15 for more information on limitations of shared savings and concerns related to federal energy financing.)

At the local level. A recent survey conducted by the American Association of School Administrators (AASA) revealed that public schools doubled their reliance on performance contracting from 1989 to 1991, and today over 3,000 districts rely on some form of guaranteed energy savings to reduce operating costs and to obtain needed equipment. The AASA survey also revealed that the large districts (over 10,000 enrollment), typically the harbingers of educational trends, intend to increase their reliance on performance contracting by 23 percent in the next three years.

For utilities. As utilities are placed under greater pressure to use integrated resource planning—and more specifically demand-side management—performance contracting has become an attractive means of satisfying these needs. (See Part II, Section 2, Chapters 16 and 17.)

At the ***state level.*** Probably the strongest endorsement for performance contracting and/or guaranteed savings is at the state level where more and more states are passing legislation to facilitate the use of performance contracting in their institutions. It began in 1985 with the passage of Ohio House Bill 264. At the end of five years under the program, the 130 participating Ohio school districts collectively showed an average savings of $75 million per year.

Performance contracting has matured into a sound industry offering a viable alternative to those who continue to waste energy—*and dollars*—because they lack the capital and/or expertise to do the work themselves.

Performance contracting guarantees savings and, therefore, results. And it guarantees satisfaction—if both parties will only take the time to understand the options and procedures, negotiate a fair contract and exercise the commitment necessary to make it work. In essence, it offers a way to use fut*ure* energy savings to upgrade your facilities or meet essential needs *today*. And it's guaranteed to work!

CHAPTER 2
COLOR US GREEN

With 20/20 hindsight, we can look back at the 1970s and see how political and economic forces separated "energy" from its broader environmental roots. The energy "crisis" of the 1970s raised the level of our consciousness of ENERGY as a single entity critical to our economy and way of life. America's great love affair with the automobile was directly threatened by lines at the gas pump. As America waited in line, the immediate and long-term availability of fossil fuels became a paramount concern. The Federal Energy Administration warned us that we had less than nine years of oil left. Our *energy* anxieties have persisted for nearly two decades.

Recently, our focus has shifted from whether we have enough fossil fuels to whether we can afford to burn what we have. "Energy" is gradually becoming part of a broader economic and environmental perspective.

Unfortunately, the 1990's rush to embrace environmental concerns now has the tail wagging the dog. Decisions about indoor air quality and clean outdoor air frequently disregard the energy and economic implications. The 1970s concern for energy as our economic life blood is being lost in an environmental frenzy. Our penchant for putting blinders on a horse and riding it into the ground has given us an economic/energy/environmental system that is seriously flawed. These flaws have lead to hidden energy subsidies and a misallocation of resources. To fix the system will require an integrated response.

THE 3 ES: ECONOMY, ENERGY AND THE ENVIRONMENT

Performance contracting provides an excellent mechanism to bring the "3 Es"—economy, energy and the environment—into appropriate juxtaposition. From its inception, performance contracting has paired energy and the economy. Companies have lived and died based on how well they assessed the economic implications of certain energy decisions.

The industry is now well positioned to bring environment into the loop. Energy service companies (ESCOs) have long recognized that their customers were not interested in "freezing-in-the-dark" savings. Every ESCO that survived the 1980s knows that its end product must be a comfortable, productive work environment.

Indoor Air

The near hysteria associated with indoor air problems in recent years has pushed the notion that indoor air problems are the direct result of the 1970s move toward "tight," energy efficient buildings. Energy "culprit" thinking has lead us to rely too heavily on ventilation as a cause and a cure. Unfortunately, it caused many to pit energy concerns against indoor environment concerns. (Chapter 6 discusses the indoor air quality/energy efficiency relationship.)

ESCOs on the cutting edge of performance contracting are now looking at ways to deliver indoor air quality *and* energy efficiency. Marrying CO_2 sensors to energy management systems is just one way ESCOs are moving to integrate the 3 Es. Such an arrangement could automatically keep the indoor air CO_2 level below 1,000 parts per million (the basis for the ASHRAE 62-1989 ventilation standard); thus, (1) saving energy, (2) operating the building more economically, and (3) constantly monitoring indoor air quality. In addition to meeting the obvious temperature and humidity needs, performance contracting can, and should, help owners and operators develop quality energy *and environmental* systems.

Getting the Green Light

The U.S. Environmental Protection Agency's Green Lights program is illustrative of the energy/environment relationship on several levels.

On a larger scale, we have the EPA in this case, not the U.S. Department of Energy, encouraging energy efficiency. The EPA Green Lights program encourages major corporations, and more recently other organizations, to adopt energy-efficient lighting as a profitable means of preventing pollution.

The EPA program also recognizes that what we do inside to become more energy efficient has a positive effect on the external environment. The Electric Power Research Institute (EPRI) has estimated that if our nation were to use high efficiency lighting to its full potential; i.e., wherever it is technically and economically justified, the following benefits would accrue:

- a 50 percent cut in electricity used for lighting in this country;
- a 10 percent drop in aggregate national electricity demand;
- a $18.6 billion reduction in ratepayer bills;
- a 4 percent cut in national CO_2 emissions;
- a 7 percent cut in national SO_2 emissions;
- a 4 percent cut in national NO_2 emissions; and
- a collateral reduction in waste associated with electrical generation; e.g., boiler ash, mine tailings, radioactive waste, etc.

Obviously, energy efficiency *inside* can have a major impact on the quality of air *outside.*

Outside Air

Our concern for outside air preceded indoor air quality concerns. More recently, global concerns regarding climate change have again shifted our focus outside and were instrumental in the Congressional passage of amendments to the Clean Air Act. As we moved in to the decade of the 90's, the Clean Air Act amendments put a new impetus behind the utilities' move to use demand-side management (DSM) as an alternative to new generating capacity. Strengthening a movement by state regulators and utilities to perform integrated resource planning (exploring and using least cost supply or demand options), the Clean Air Act offered utilities the carrot and the stick to reduce emissions through energy conservation. As discussed in Chapters 16 and 17, the DSM movement is providing the performance contracting industry a new and rapidly expanding market.

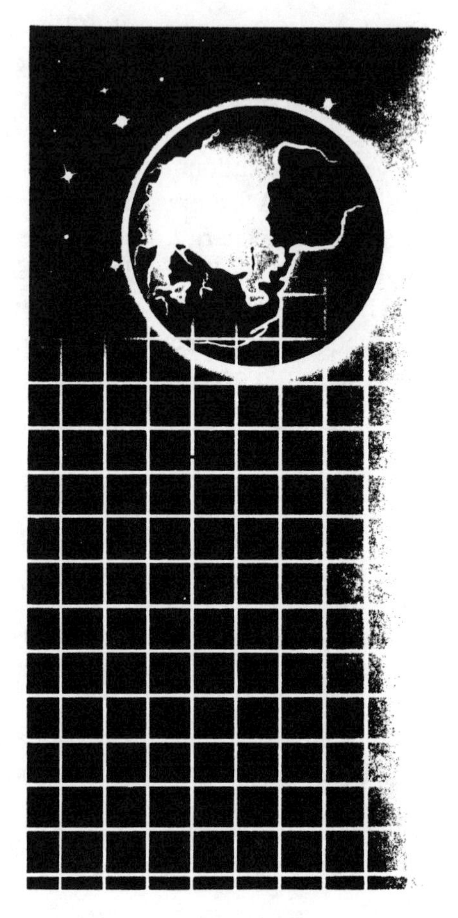

The economic and environmental benefits that energy efficiency measures, and particularly performance contracting, can deliver are not limited to the United States. The performance contracting concept will be particularly valuable and advantageous as Eastern Europe moves into a free economy and seeks to retool to a higher level of energy efficiency and environmentally conscious operations.

Externalities

The importance of including externalities in the cost of energy undergirds much of the movement toward integrated resource planning and is a critical aspect in implementing the Clean Air Act.

Putting a price tag on the societal value of reduced impacts on human health and the environment is not easy. In effect, some dollar approximation needs to be assigned to the way production and distribution of fossil fuels effects our land, water and air.

The cost of *assigning the value* of externalities to fossil fuels and generated electricity, in and of itself, will add to the fuel bill. The added costs will vary depending upon the point in the production chain where responsibility and accountability for pollution and control are assigned. Facilities, for example, that now accept oil from 15 companies into one or two holding tanks may need 15 tanks if regulations require that problems be traced to a specific oil company.

To be realistic the price of energy should also absorb the cost of protecting our "vital interests" in keeping peace in the Middle East. Our current dependency on foreign oil, gives OPEC and individual sovereigns enormous control over our economy and our daily lives. Protecting our sources and assuring availability of oil at a reasonable price does not come cheap.

Ed Stein's cartoon from the *Rocky Mountain News* may exaggerate the cost but it does make the point that protecting our vital interests does not come cheap.

As the real price of energy is uncovered, alternative energy sources and energy conservation will become increasingly attractive. Fortunately, savings in one area can reduce costs for all 3 Es. For example, preliminary calculations by the U.S. Department of Energy indicate that an expenditure of $50 billion on converting all swimming pool/spa/hot tubs to solar would result in significant savings which annually would offer:

- a return on investment (ROI) of 33 percent;
- an energy savings of 5.2 quads; and
- reduced CO_2 emissions of 340 million tons.

Potential Impact

$50 billion per year spent on swimming pool/spa/hot tub heating

5.2 quads of energy per year used for swimming pool/spa/ hot tub heating

340 million tons of CO_2 emitted per year in heating swimming pools/spas/hot tubs

Figure 2-1.
Potential impact of "solarizing" the nation's pools/spas/hot tubs.

Economic advantage - Aside from a healthy jolt to the economy that could be generated through $50 billion in conversions, the 33 percent ROI offers a very attractive investment.

Energy advantage - Converting 5.2 quads into terms we can understand, Figure 2-2 shows the energy equivalent of 5.2 quads.

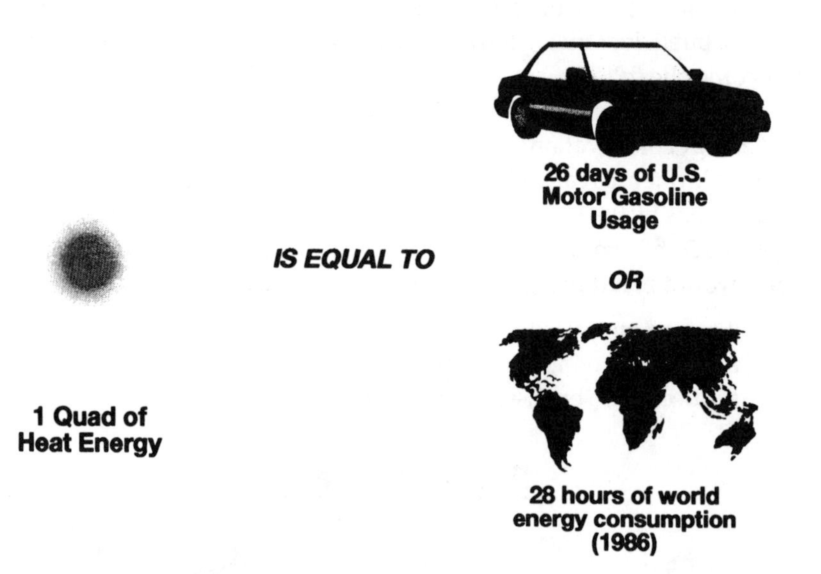

Figure 2-2. Energy equivalent of a quad.

Environmental advantage - The displacement of 340 million tons of carbon dioxide is equivalent to planting 224 million trees according to U.S. DOE.

In addition to the CO_2 reduction, U.S. DOE preliminary calculations show solarizing the heated, non-solar swimming pools/spas/hot tubs would also save each year:

- one million tons of NO_X;
- 600,000 tons of SO_2; and
- externality costs of $8.2 billion

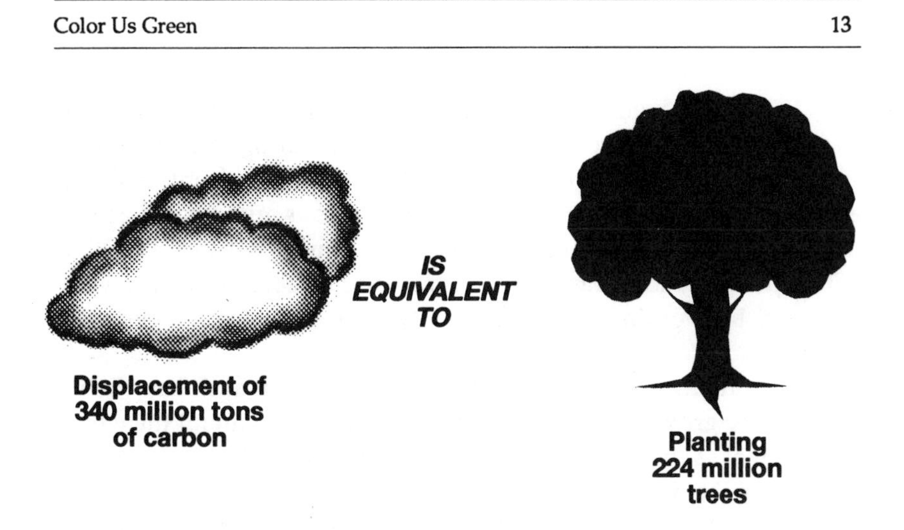

Figure 2-3. CO_2 equivalent in trees.

It would appear that solarizing swimming pools saves money, energy and reduces environmental damage. The catch, of course, is finding the initial $50 billion in capital to do the job and the expertise to be sure it is done right. The U.S. economy in the early 1990s has not left any spare change on the table. Certainly not $50 billion!!

Making It Happen: Performance Contracting

Enter performance contracting stage left. If DOE's initial numbers hold up in actual practice, energy service companies can solarize as fast as the product can be manufactured. The performance contracting potential, however, must be tempered by administrative/economic feasibility. ESCOs can't afford to do each and every hot tub on a separate performance contract, but in concert with a utility in a DSM program, it might become a viable market.

Performance contracting is an ideal vehicle to integrate measures that save dollars, reduce energy consumption and avoid damage to the environment. A pragmatic industry, performance contracting offers an ideal vehicle to merge our separate "3 E" concerns into a viable national program.

If you have read this far because you are looking for performance contracting services, you are already sensitive to the economic and energy aspects. If you were to assess the situation with internal and external environmental concerns in the mix, your arguments become much more persuasive and the opportunity much greater. Working with a performance contractor, you may also be able to enhance your project(s) by working with your utility to help it meet its clean air objectives.

Kermit has reminded us over the years that "It ain't Easy Being Green." The growing environmental conscience throughout the world not only makes "being green" easier, but a necessity. In fact, in these days a company that isn't green, is apt to end up in the red.

Recognizing the capacity they have to serve several masters, energy service companies must learn to determine the positive environmental impact their actions can have. When a measure saves energy, it saves on noxious emissions. This benefits us all. Green marketing should be high on every ESCO's agenda.

CHAPTER 3

THE BOTTOM LINE

In performance contracting, what a proposed modification will cost and save is the bottom line. Return on investment is usually the motivating force for both the end-user and the energy service company (ESCO).

Conversely, the dollars that will be lost if you don't act should be a major factor in setting financial priorities. You must, therefore, consider how much your organization will spend if it doesn't reduce energy consumption. This analysis should precede decisions related to energy efficiency and are integral to effectively managing a performance contract once it is in place.

More than one manager has been furious at an ESCO because the organization wasn't achieving predicted "savings," when, in fact, energy savings had been eaten up by changes in the rate schedule. The "front office" may not understand how changing rate schedules can destroy apparent savings. The concept of cost avoidance is essential to weighing and communicating performance contracting benefits.

The guidelines for weighing cost-effectiveness, calculating cost of delay and computing/graphing cost avoidance offered in this chapter are important to you, the end-user, as well as to ESCOs.

Cost-Effectiveness

Because every business, every organization uses a significant amount of costly energy, a wide array of energy conservation opportunities are available to them. However, these opportunities must be weighed to determine which measures offer the greatest financial benefit. This requires not only an evaluation of the cost-effectiveness of the measures, individually and in combination, but a broader financial analysis as well.

Ideally, major modifications will be reviewed by a committee constituted by management so that implications for particular programs or impact on the work environment are part of the decision making process. Indoor air quality and other environmental concerns should be considered. At the very least, a cost/benefit analysis requires the sharing of information and techniques between the organization's business officer and the facility manager. If your organization has an energy manager, he/she should definitely be involved. The group should collectively weigh all the energy conservation measures in the light of the financial benefits to the energy program and to the total operation.

Other operation and facility considerations also need to be considered. A new roof is not apt to be the most cost-effective measure. However, if the old roof is leaking, a new roof with increased insulation may be the most critical need. Or, replacing an old boiler that is not only inefficient, but is unpredictable and demands a lot of maintenance, may take precedence over a more attractive controls option.

Sources of funding and reimbursement implications must also be considered. If the local utility has a rebate program to encourage replacement by more efficient lamps, for example, lighting changes become even more attractive.

Cost-effectiveness is one measure of economic feasibility. It is an essential ingredient in performance contracting. It answers the question: "How soon can we get our money back from this investment?" There are various ways to calculate the time necessary to recoup the cost of the original investment. These range from simple payback and adjusted payback to the more complicated life-cycle costing (LCC).

Simple Payback

Quick, simple and universally understood, simple payback calculations generally provide sufficient data for low to modest investments. It can also provide a good "first cut" on larger investments. The simple payback period (SPP) is found by dividing the value of the initial investment by the projected annual energy savings. SPP is usually given in years and/or tenths of a year.

Simple Payback Formula

$$\text{SPP (years)} = \frac{\text{I}}{\text{ES/year}}$$

where,

SPP	= Simple payback period
I	= Initial investment
ES/year	= Projected annual energy savings at current prices

if,

SPP	= $ 2,000
ES/year	= $ 450

then,

$$\text{SPP} = \frac{\text{I}}{\text{ES/year}} = \frac{\$2{,}000}{\$450} = 4.4 \text{ years}$$

Adjusted Paybacks

The simple payback calculations may be modified by any factor management finds critical. The more common adjustments to be factored in are; (1) changes in operations and maintenance (O&M) costs, or (2) projected changes in energy costs.

1. Changes in O&M costs. The new equipment might require more, or less, O&M work. To the extent the O&M work can be quantified, the Adjusted Payback Period (APP) formula would then read as follows:

Adjusted Payback Formula, O&M

$$APP_{O\&M} = \frac{I_n}{ES_n \quad \pm M_n}$$

where,

$APP_{O\&M}$	= Payback period adjusted for O&M
I_n	= Initial investment for period of analysis
ES_n	= Energy Savings for analysis period
M_n	= Differential operations and maintenance costs for analysis period
n	= Period of analysis

This approach is sometimes considered in terms of Simplified Cash Flow (SCF). In this case, the computation is usually calculated within the parameters of a given fiscal year. SCF weighs the difference in the cost of the fuel consumed plus the difference in O&M costs against the investment for a given time period.

Simplified Cash Flow

$$SCF = (E_n + O\&M_n) - (I_n)$$

where,

SCF	= Simplified cash flow
E_n	= Energy cost savings for the time period
$O\&M_n$	= Operations and maintenance savings for the period
I_n	= Initial investment prorated
n	= Period of analysis

if,

I	= \$2,000, spread over 4 years
E_n	= \$450 per year
$O\&M_n$	= \$150 per year

then,

SCF = (\$450 + \$150) – (\$2000/4)
= \$600 – \$500 = \$100/year

2. Changes in energy costs. Since the payback period is a function of energy costs, this approach can more accurately reflect the impact of unstable energy prices. The difficulty comes in predicting future energy costs. Therefore, it is recommended that you limit these projections to three years and use them *only for internal discussion*.

Adjusted Payback Formula, Energy Cost

$$APP_e = \frac{I}{X\left(E_{n_1} + E_{n_2} + E_{n_3}\right)}$$

where,

APP_e = Payback period adjusted for projected energy costs
I = Initial investment
E_{n_1} = Projected 1st year energy costs

E_{n_2} = Projected 2nd year energy costs

E_{n_3} = Projected 3rd year energy costs

It is worth restating the inherent dangers in publicly justifying certain energy conservation measures by predicted energy price increases (unless you are quoting data supplied by the utility). If prices don't increase as predicted; or, worse yet, fall, you end up looking pretty bad. Even worse, the justification may become the story. The headline resulting from a university's board of trustees meeting may state, "20% Energy Price Increase Next Year: University Says." And the planned energy efficiency work becomes lost in the hullabaloo.

Life-Cycle Costing

Incorporating all costs and savings associated with a purchase for the life of the equipment is increasingly being used as a means of judging cost-effectiveness. This approach, Life-Cycle Costing (LCC), may appear to administrators in government to be the antithesis of the required low bid/first cost procurement procedures. If specifications call for LCC as a means of determining cost-effectiveness, then LCC can be compatible with low bid procedures.

LCC's rather rigorous approach can be quite time consuming; however, you will find the effort is usually justified for larger purchases and/or for relatively limited capital. Life cycle costing addresses many factors which an adjusted payback analysis may miss—salvage value, equipment life, lost opportunity costs for alternate use of the money, taxes, interest, and other factors.

The simplest mode of analysis for LCC is:

Life-Cycle Cost Formula

$$LCC = I - S + M + R + E$$

where,

LCC = Life-Cycle Cost
I = Investment costs
S = Salvage value
M = Maintenance costs
R = Replacement costs
E = Energy costs

LCC is the net benefit of all major costs and savings for the life of the equipment *discounted to present value.* A building design or system that lowers the LCC without loss in performance can generally be held to be more cost-effective. Other considerations, such as the calculation of present worth, discounting factors and rates and LCC in new design, need detailed analysis and are more fully discussed in *Life-Cycle Cost Manual for the Federal Energy Management Program,* prepared by the National Bureau of Standards, United States Department of Commerce, and *Introduction to Life Cycle Costing,* published by the Fairmont Press.

Cost of Delay

Some things may be put off without a loss of revenue. Energy efficiency work cannot. Every tick of the clock, every day that passes, represents dollars your organization may have wasted by consuming needless energy. Every hour of delay forces you to give to the utility money that, through energy management, could have been used to educate students, train sales reps, offer patients additional services, launch a media campaign, meet constituent needs, make a bigger profit, etc. Many administrators tend to treat the utility bill as an inevitable cost. Others find that organizational pressures requiring immediate attention push energy concerns aside. But, every day you don't act is a day of wasted energy and, what is more important, ***a day of wasted money.***

Those working in energy management have become accustomed to weighing options by calculating cost-effectiveness as discussed above. The rapidity with which energy savings recover initial investments should be a major factor in weighing energy retrofit *vis a vis* other investments. It is not unusual to find a school district, for example, with a reserve fund earning 7.1 percent interest while energy investments with an ROI of 25 percent or more go begging. You and others in your organization need to fully understand the "earning potential" in energy efficiency and the lost revenue inherent in cost of delay.

The cost of delay is almost the mirror image of the cost-effectiveness profile. The same factors that contribute to energy cost/benefit analysis affect cost of delay calculations, but in a negative sense.

One means of calculating cost-effectiveness, Simplified Cash Flow (SCF), as discussed above has an especially good fit. To repeat, the SCF formula is:

$$SCF = (E_n + O{+}M_n) - (I_n)$$

where,

SCF	= Simplified cash flow
E_n	= Energy cost savings for the period
$O\&M_n$	= Operations and maintenance savings for time period
I_n	= Initial investment prorated
n	= Period of analysis

By totaling the differential costs (the anticipated savings) and subtracting the prorated investment, the positive cash flow for a specific cost-effective energy conservation measure is determined.

The same formula can be used for Cost of Delay (CoD) where the lost savings potential becomes the differential. The lost savings differential is then reduced by the outlay that would have been needed to achieve those savings, the prorated investment. This negative cash flow figure represents the cost of postponing energy work.

$$CoD = -(E_n + O{+}M_n) + (I_n)$$

where,

CoD	= Cost of Delay
E_n	= Energy cost savings for period
$O\&M_n$	= Operations and maintenance savings for period
I_n	= Initial investment prorated
n	= A specified period of time

For example, suppose the city government is just starting an energy program and has been advised it can reduce consumption by 25 percent through energy efficient O&M procedures, low-cost retrofits, and the installation of an automated energy management control system. With an annual utility bill of $1.6 million, the avoided costs could be $400,000 per year, less the cost of work. If a $1,400,000 installation prorated over five years could save $400,000 per year in energy and operations and maintenance costs, the SCF would be $120, 000 per year. On the other hand, no action represents a CoD of *minus* $120, 000 per year.

Using an annual SCF formula without any adjustment for inflation and with the purchase/installation costs prorated over five years, the calculations would be:

SCF = (E + O&M savings for year) – (I [prorated])
= ($400,000) – ($1,400,000/5)
= $400, 000 – $280, 000
= + $120, 000

The $400,000 savings would reduce the existing $1.6 M to $1.2 M. In this case, the potentially lower budget is used as a reference point.

CoD = – (E + O&M for the year) + (I prorated)
CoD = The new lower budget minus the existing budget
($1. 2 – $1. 6 = – $400, 000)
Plus the prorated investment
($1, 400, 000/5 = $280, 000)
= – $400, 000 + $280, 000
= – $120, 000

James F. Whiteman

In essence, this rather awkward formula looks at what could have been saved less the cost of the work required to achieve those savings over a specified period. To put it more simply, potential savings equal your losses if you do nothing.

The formula does not look beyond the years of prorated investment. In the above example, the CoD becomes $400,000 (exclusive of any O&M costs, depreciation or net present value adjustments) in the sixth year and every year thereafter, for the life of the improvement.

Even when using internal resources, you should calculate the Cost of Delay related to various financing options. Using limited resources for other organizational needs may, in the long run, prove to be more costly. If those needs are paramount and the budget is tight, then getting the work done through performance contracting becomes an even more viable option. Figure 3-1 compares the cost of delay a school district might incur by doing nothing to the financial benefits of using traditional bonding or performance contracting. The figure uses simple payback

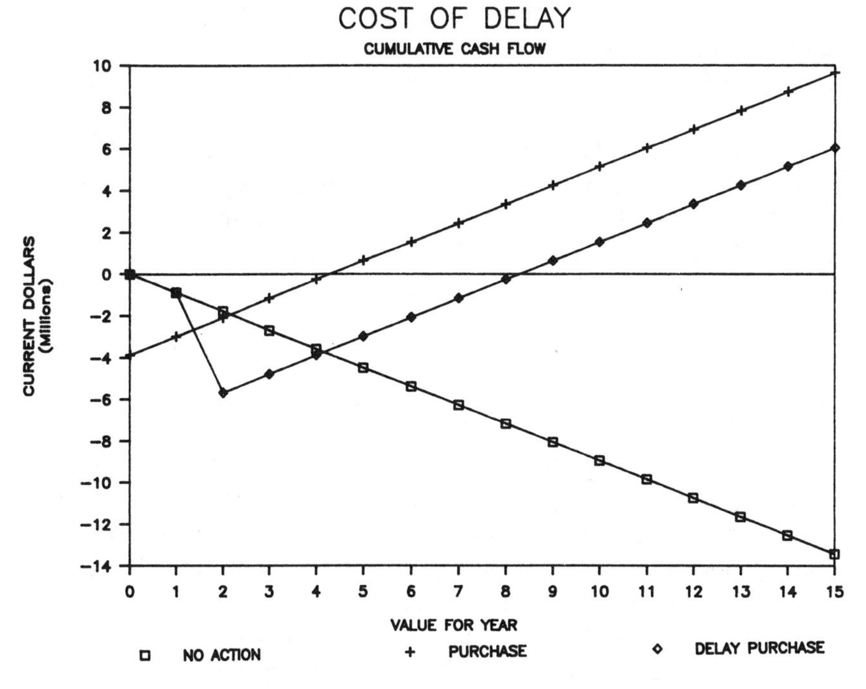

Figure 3-1. Cost of delay.

and constant dollars.

An organization is frequently reluctant to use performance contracting because it prefers to do the work itself and "save the service costs." *Reality seldom meets expectations.* Frequently, the losses due to delays exceed the benefits of internal funding. When you are considering the use of in-house funds to save money, you should carefully weigh the Cost of Delay first. And the delayed period should be based on how long it took to get similar work in place in the past—from conception to acceptance of the installation—and not what you hope might be the case this time.

In addition to the cash flow impact depicted in Figure 3-1, you need to also recognize the benefit in improving your organization's capital stock, which ultimately enhances its fiscal condition and usually reduces operations and maintenance costs.

To check your understanding of the calculations needed to determine the annualized Cost of Delay you may wish to work through the problem in Figure 3-2.

Using data supplied through energy audits, you can use the same Cost of Delay calculation procedures to determine what you may actually be losing every day, month, or year that your organization fails to act.

After the commitment to secure private sector financing has been made, delays are often incurred because outside support, particularly attorneys, are not familiar with the performance contracting process. These delays can be reduced if your attorney and engineer are involved early in the process.

Cost Avoidance

Rising prices can wipe out all the dollar gains you have made by reducing energy consumption. But think what your organization would have been paying if it hadn't cut back! In order to communicate energy management benefits to others as costs rise, its important to be able to talk about what would have been—the costs avoided.

The joy of counting the dollars that would have gone to the utility makes cost avoidance very real and very gratifying. In order to calculate cost avoidance, a base year must be established. The base year consump-

Cost of Delay: Sample Problem

Data:

Annual utility bill	=	$1,500,000
O&M on existing equipment	=	600
Projected savings	=	30% energy
		30% O&M
Retrofit investment	=	$1,750,000
Prorated period	=	5 years

Problem:

If the project is put off for one year, what is the Cost of Delay?

First, calculate and projected savings

$$\$1{,}500{,}000{,}000 \times 30\% + \$600 \times 30\%$$

Then, determine what it would cost each year to achieve those savings

$$\$1{,}750{,}000 \div 5 \text{ years}$$

(The net LOSS should be – $100,600/yr)

Now, calculate CoD for 6 years (without considering depreciation, time value of money, or adjustments in O&M.)

Answer: – $953,000

($100,600 × 5 + $450,000 in 6th year, after investment is paid off.)

Figure 3-2. Cost of delay analysis.

tion multiplied by the *current* price per unit will reveal "what it would have cost."

Cost avoidance is what it would have cost minus current costs. Most top management or board members seldom have the time or inclination to wade through a pile of numbers; so it will pay to graph the data when cost avoidance for more than one year is involved.

As an illustration, Figure 3-3 depicts a cost avoidance analysis for Tender Care Hospital. The hospital had cut electrical consumption by 1,002,660 kWh since 1984, but experienced a rate increase of $0.042/kWh over the 1984 to 1992 time period. Even though con-sumption dropped 25 percent during this time period, costs still rose about $73,000. Without energy efficiency measures the cost would have increased $167,000. The top line in Figure 3-3 depicts what it would have cost if consumption had remained at the 1984 level, the middle line indicates actual costs. The shaded area between the top and middle line shows the avoided costs. The bottom line shows the decline in consumption.

The same type of graph as shown on the following page could be used for all the fuels used in a building if total Btu is placed on the vertical axis.

Cost-effectiveness, Cost of Delay and *Cost Avoidance* are critical components of energy decision-making whether you plan to do your own retrofit work, or ask a performance contractor to provide the service.

ESCOs need to employ these concepts as part of their marketing technique. Cost avoidance numbers, preferably graphs, should be a regular feature in the monthly billing process. Without it, customers quickly forget what the utility bill used to be and why they are paying the ESCO. This becomes particularly critical when there has been a change in administration.

PLOTTING COST AVOIDANCE

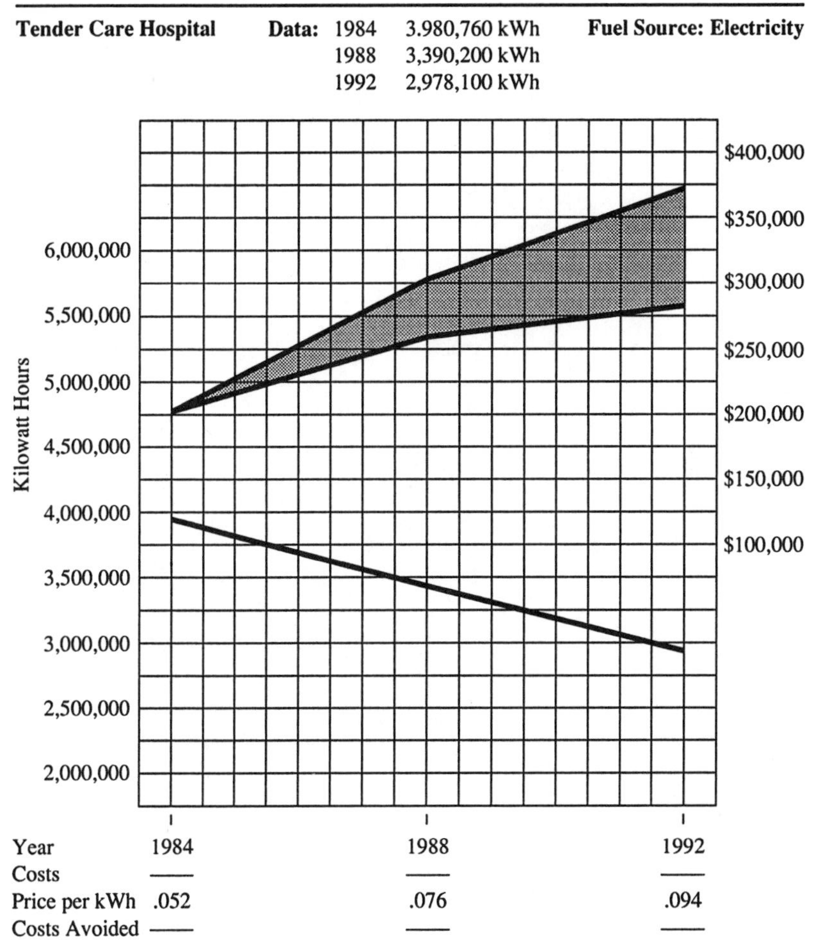

Figure 3-3. Cost avoidance graph.

CHAPTER 4

ADMINISTERING A PERFORMANCE CONTRACT

Performance contracting lives or dies on the sense of partnership achieved. Those energy service companies (ESCOs) that survived the 1980s recognize a strong partnership as a critical component of any effective savings-based agreement. The level of commitment that is exhibited by management, therefore, becomes a decisive factor in whether or not an ESCO will invest in a certain business, university or hospital.

Performance contracting requires effective energy leadership by both parties. Exercising that leadership in the customer's organization is usually more complex than in the ESCO, primarily due to the fact that it usually involves many people, who do not have energy concerns as a major responsibility.

GETTING THE MOST OUT OF PERFORMANCE CONTRACTING: THE CUSTOMER

A tremendous advantage to an organization that elects to use performance contracting is the expertise it gains from the ESCO backed by its guaranteed performance. Unfortunately, this advantage can be so tantalizing that some organizations assume the ESCO can do it all; and, in effect, they abrogate their own responsibilities to the effort.

The ESCO can bring its expertise to the partnership, but occupant behavior can make all the difference in the level of savings. Uncommitted administrators, uncaring staff and indifferent operations and maintenance personnel can subvert or undermine the best an ESCO has to offer.

ESCOs do not guarantee the maximum savings, the best that can be achieved under ideal conditions. Understandably, the ESCOs have a vulnerability that causes them to be cautious with their guarantees. Guaranteed savings is apt to be around 80 percent of what an ESCO can *reasonably* project the savings to be. An organization that does not work hand-in-glove with its ESCO can impede the ESCO from reaching even the 80 percent mark. But with a strong supportive partnership, both parties can enjoy a positive cash flow from the other 20 percent or more.

Putting the "manage" in energy management requires both administrative commitment and leadership. The real difference between an effective energy program—and one that is not—is the attention paid to the human dimension. It is up to you and the rest of management to mobilize this potential support. Energy leadership qualities draw from the array of management skills already in hand and redefines them within the context of the organization's energy concerns and opportunities.

For most managers, limiting the work day to eight hours sounds like a remote dream. The mere thought of adding to the myriad of administrative responsibilities is enough to make most managers shudder.

But energy is not *added* to the day. It permeates all activities. An effective program requires that energy considerations be squeezed in between finding the money to buy new stadium lights, learning the latest "red bag" procedures, filling out government forms, planning an ad campaign or getting a report ready for the commissioner. Fortunately, once its components have been set in motion, a comprehensive energy management program under ESCO monitoring usually requires only limited attention and reassessment. Visible and continuing attention, however, is essential.

The binding force in energy management and performance contracting is communications. A key skill in managing energy use is the ability to communicate energy needs and opportunities and to elicit the support needed to make the human dimension a strong core of the energy program. This requires continuing attention to the communications function by each partner, both internally and between the partners. Communications is one aspect of a performance contract, or any effective energy program, that costs little and gains much.

One of the greatest ironies in energy management is that planned communication can be the most cost-effective tool in energy management; yet, it is often relegated to an incidental, haphazard approach. With the result that communication strategies are too often missing, insufficient or misplaced.

Experienced ESCOs not only survey the energy saving potential of a building, but of the administration and staff. The most attractive customers have a written energy policy in place, a designated person responsible for energy efficiency and a demonstrable commitment to both.

Organization Policy

Administrations can be transitory. Adopting a board policy cements energy positions, gives enduring guidance and reassures the ESCO that support is stable. Frequently, the support for performance contracting is part of the organization's energy policy, integral to a broad statement of administrative commitment to effective energy management.

Typical policy statements include:

- A statement of concern regarding the broad energy situation and more particularly economic and supply implications for the organization;
- A short statement recognizing the advisability and cost-effectiveness of developing energy management procedures;
- A statement of commitment;
- Preliminary implementation considerations such as;
  —authorizing the position of energy manager;
  —delegating authority to the energy manager within specified parameters;
  —requesting that an energy management plan be submitted for board and/or administrative approval; the plan to include goals (potential reductions), energy costs (history patterns/projections), potential savings, suggested procedures, recommendations, budget, and ***potential funding procedures*** [performance contracting may be explicitly mentioned];
- Reporting requirements, which would incorporate evaluative data and further recommendations.

A board policy should be just that, a brief statement of policy. It should not include day-to-day considerations, or the mechanics of implementation. Specific implementation plans, such as temperature settings, belong in an energy management plan. Depending on your organization's operational procedures, the plan may or may not be approved by a board.

The Energy Manager

If a person has not been designated as an energy manager, even for a portion of his or her time, then no one is managing energy use.

At the very least, one person should be designated as responsible for energy matters and given the time, the authority, and the budget to do the job.

If an organization is retaining an ESCO to do its energy worrying, does it need an energy manager? That depends on how comprehensive the contract is. At the very least, part of someone's time needs to be devoted to working with the ESCO and making sure your organization and the ESCO are both holding up their end of the partnership.

Key ingredients of an energy manager's job. Although the job description needs to be situation specific, it should include:

- setting up and/or implementing an energy management plan;
- establishing and maintaining energy records by consumption and cost;
- identifying rebates or assistance available from other sources; e.g., utilities, federal grants, and exploring ways to leverage such funds or assistance;
- assessing energy needs; overseeing energy audits;
- identify sources of energy financing and weighing the relative financial benefit for various financing options; soliciting performance contracting proposals and evaluating ESCO qualifications;
- making energy recommendations in conjunction with the energy committee based on such criteria as the prevailing codes, feasibility, cost-effectiveness, financial benefit, health care and safety needs, and maintaining the work environment;
- implementing approved recommendations: writing specification, overseeing procurement, installation, fine tuning, and operation;
- serving as a liaison to energy committee and contact point for the performance contractor;
- planning and implementing internal and/or external communications strategies, or supplying energy information to those responsible for that function; and

- evaluating the program's effectiveness, updating it, and routinely reporting progress to management and/or the board.

Qualifications an energy manager should have. An effective energy manager needs to have technical expertise and some financial insights as well as communications and leadership skills. Unfortunately, persons with technical expertise do not always have the verbal skills required. The available choices range from a truly fine engineer with limited communications skills to the eloquent leader, who doesn't know a Btu from a cup of coffee.

Given a choice, the central range with balance in technical and verbal skills is desirable. The ultimate answer obviously varies depending on what is needed, what engineering or communications capabilities are already on staff, and the relative importance attached to hardware vs. user considerations.

The energy manager must be in close and constant communication with the people who pay the bills. A brief session involving the financial officer, the energy manager, and the utility representative to assess demand profiles and their implications for operation can often reduce energy cost by thousands of dollars. This can, and should, be done before a performance contract is entered into, so these easily secured savings do not have to be shared with an ESCO. If it is not already done when the ESCO enters the process, then the ESCO should be asked to facilitate the process and be a party to it. You can still keep any savings earned by understanding the purchasing process.

To elicit support from the building occupants, the energy manager must be able to articulate energy needs and conservation benefits in a fashion that strikes a responsive chord in the users. Generally, this requires overcoming the energy rhetoric fatigue and skepticism resulting from being inundated over the years with "crisis" reports and conflicting data. Strategies to meet this problem are addressed in Chapter 13. It is important to note here, however, that the responsibility to implement these strategies rests with the energy manager.

Who should an energy manager report to? Experience has shown that in order to effect change, an energy manager must hold a position of stature within the organization's hierarchy. While situations will vary with personalities and local conditions, it is generally advisable to have the energy manager report directly to top management; e.g., the hospital administrator, school business official, financial officer, superintendent or president. Some factors to consider in making this decision are:

- Decisions about operations (air exchanges and circulation containment needs, class schedules, athletic events, etc.) frequently necessitate discussions with directors, principals, department heads, deans, medical staff or sales managers, who all have their own worries and goals. Such discussions must be held among individuals with approximately the same stature. *It is hard to effect change from a position buried in the facilities department.*

- Implementing an energy program almost always requires changes in past practices. It is not unusual to find resistance to these changes. If it is the boss' past practices that need changing, making a difference can be exceedingly difficult. If the energy manager is to "make a difference" in operations or maintenance, he or she must be on equal footing with the director of facilities, chief engineer, etc.—not reporting to him or her.

- The relative position the energy manager holds in the organization is regarded as an expression of the emphasis management places on the energy program and its commitment to it.

In short, for your organization to have an energy program that makes a difference, the energy manager's position must be high enough on the totem pole to get the job done.

Project Assessment

In the early days of performance contracting, the merits of "shared savings" used to be billed as "Their Cash; Your Savings." Like a complimentary dinner guest in a restaurant, the guest doesn't worry abut the size of the check. Through the years, however, has come the realization that you may not pay up front, but at some point you will pay. The funds may be "off balance sheet," money that previously went to the utility, but no matter how you slice it: ULTIMATELY IT COMES OUT OF YOUR POCKET.

As in all things, you should assess what you are getting for your money before signing the contract, during the installment phase and throughout the life of the contract. If your organization doesn't have the how-to-do-it resources, in-house, they are available outside. You need not be impeded by the front end cost of such services; for, unless they become too exotic, they can all be assigned to the project. Outside resources may include:

- an engineer with energy conservation credentials (not an architect unless the training or experience includes mechanics, electricity and energy efficiency);

- a contract attorney with performance contracting experience;

- a financial consultant with some understanding of life cycle costing and energy efficiency.

- an independent assessment group to do periodic checks on the formula, the adjustments to it and the savings calculation; and

- a performance contracting consultant, who can look out for the organization's best interests, guide it through the process and put the organization in touch with top notch people who provide the above services.

Effective project management also demands a means of independently assessing the savings achieved. Some computer-based energy accounting program, such as FASER from Omni Comp, should be an integral part of your project assessment practices.

In summary, the inescapable components of an energy program that make a difference in the bottom line are a statement of commitment, designated responsibility, a plan, an identification of the resources to do the job, on-going assessment capabilities and a carefully designed communications strategy.

The importance of administrative leadership and the inherent communications functions can best be illustrated by considering the energy audit. The audit is a valuable tool, but audits don't save energy; people do. The unattended audit report gathers dust. Only when it is read, discussed, *and implemented* can its energy dollar benefits be realized. The difference between dust and energy savings is people. It is the communications connection that makes an energy program work.

ESCO MANAGEMENT STRATEGIES

The hit and run approach doesn't work. ***Performance contracting requires performance over the life of the contract.*** ESCOs that thought, once a sale was complete, all they had to do was perform a little maintenance and issue a monthly bill are no longer with us. The industry is full of horror stories that can be traced to poor ESCO leadership and poor communications.

A working partnership requires more than lip service. The effective ESCO looks at the organization's needs and *serves them.*

Service that is directly related to savings cannot be farmed out. For example, one ESCO in the 1980s contracted out all of its monitoring, billing and maintenance on existing contracts for a set fee. The ESCO lost contact with its customers. The subcontractor had no incentive to provide the quality of service that meant greater savings. Everybody lost.

Performance contracting should be a way for an organization to unload its energy worries, its equipment needs and its energy-related maintenance headaches.

The organization, however, should not assign more nor expect more, of its ESCO partner than the ESCO has historically offered its previous customers. Grand and glorious promises, which the ESCO does not have the resources to deliver, paves the way for growing disenchantment that can ultimately lead to court.

The best performance contracts today are much more than financing. They offer specialized energy expertise, improved capital stock, training, monitoring—whatever the customer wants that can feasibly be built into the package. In fact, the effective ESCOs are increasingly laying out a smorgasbord of opportunities from which customers can pick and choose services that precisely meet their needs. And those needs may not even include financing. They may, however, include guarantees that enhance the customer's ability to do its own financing.

Most aspects of an effective ESCO's operation are common to any well-run organization; however, energy service companies have a few peculiarities that require special management sensitivities and procedures.

First, ESCOs sell a service, but customers usually think they are buying products. On one hand, performance contractors sell cost-effective productive environments: the boilers, chillers and controls that may be installed are merely vehicles to make it happen. The hospital, on the other hand, needs a new boiler. From the administration's perspective, getting a more efficient boiler through performance contracting, means the boiler is paid for from future energy savings.

This difference in point of view can undermine effective communication between the two organizations and frustrate the sense of partnership.

Second, the ESCO is selling promises; predictions of savings. When those savings are realized everyone is happy. But memories are short. The high utility bills of the past may be forgotten, especially if there has been a change in management, and soon the customer may feel he's paying for "nothing."

One important ingredient ESCOs can, and should, provide their customers is cost avoidance information. A performance contract is designed to save energy, not money. Performance contracts sold as "money savers" become management headaches if the price of fuel goes up. ESCOs, who want to develop and maintain good relationships with their customers, learn how to calculate cost avoidance and to communicate those "savings" graphically with billing procedures: or, at the very least, annual reports.

Third, to the uninitiated the process seems to rely on smoke and mirrors. It requires some expertise in both financing *and* energy technology if the customer is to be comfortable with the performance contracting concept. Bridging the gap from business office to boiler room is not always easy.

The performance contracting process also requires an attorney, who knows something about contract law, financing and energy efficient technology. Too often, rather than acknowledge their limitations, lawyers drag out the decision-making process or advise against the whole concept. ESCO sales personnel and management must be all things to all people, or draw on *independent* third party financial, legal and technical contacts to educate their customer's counterparts.

The good ESCO anticipates what the energy manager needs for him or her to maintain an informed management and staff. Then, makes sure he or she has the information when it's needed. Such an effort not only maintains commitment and enhances savings; it paves the way for contract enhancement and renewal.

CHAPTER 5

MAKING CENTS OF BTU

How do you decide what energy efficiency measures to take? How do you know when they work? How can you gauge the effectiveness of the energy service company's work?

Comparing utility bills with the same month last year gives us some idea of how we are doing until the utility changes its rate schedule. More than one facility manager's job has been on the line because he or she claimed improved energy efficiency; but, having tracked only costs, the manager could not document any savings because the utility had raised the rates.

Btu are what you buy every time you pay the electricity, gas, and/or oil bill. The more Btu you use; the higher the bill. The more efficiently a facility uses energy; i.e., the more work it gets from the energy it buys; the lower the Btu—*and the costs.*

BTU = Dollars

Understanding how Btu are bought and used is as important to the business office as it is to the boiler room.

Organizations *spend* dollars for energy, but they *consume* Btu. To save the dollars, it's important for you to know how the Btu are consumed, how consumption can be reduced through more efficient operations, and how to keep track of the progress being made. The British thermal unit (Btu) offers a way to measure energy consumption no matter what source is used. Or, to add sources together to determine the total energy required to operate a facility.

MANAGEMENT'S GUIDE TO THE WONDERFUL WORLD OF BTU

Maintaining records only by tons of coal, gallons of oil, or kilowatt hours is only partly useful because these units cannot be added together and used to calculate a building's total energy consumption. And *total* energy consumption dictates the bottom line. Changing to more efficient light bulbs saves electricity, but the loss of heat due to the more efficient bulbs increases the need for heating fuel. Only when electrical and heating fuel consumption is added together is the real savings known. Therefore, to make these calculations and to compare existing energy consumption figures with pre-modification consumption you will need to find a common denominator, the British thermal unit (Btu). A Btu is the unit of heat energy necessary to raise the temperature of one pound of water one degree Fahrenheit at sea level.

Conversion Factors

Each fuel source has a certain number of Btu per unit. To combine units of various fuels for calculation, they must first be converted to Btu. Table 5-1 indicates the conversion factors for changing units of fuel or electricity to Btu.

If a facility is heated with natural gas, then the total Btu consumption can be calculated by multiplying the thousand cubic feet (mcf) consumed during a certain period of time by 1,030,000 and adding it to the product of kilowatt hours (kWh) for the same period times 3, 413. The resulting Btu provides a measure of the total energy consumed during the specified time period. For example, Central High School used 5,760 mcf of natural gas and 3,410,000 kWh of electricity during the 1990-1991 school year. To find total energy consumption for that school year, we would make the following calculation: (5, 760 mcf X 1, 030, 000 Btu) + (3,410,000 kWh X 3,413 Btu) = total Btu consumed, or 17,571,130,000 Btu.

Because numbers this size become rather cumbersome, they are usually converted to million Btu (MMBtu). In this case, 17. 57 MMBtu.

To compare Central High School's consumption to smaller Lincoln Middle School, it is necessary to eliminate the size factor by dividing the total Btu consumed by the conditioned (heated and/or cooled) gross square feet in the building, or:

$$\text{Btu/square foot} = \frac{\text{units of fuel} \times \text{conversion factor}}{\text{gross square feet}}$$

Btu/square feet (sq.ft.) is often referred to as an Energy Utilization Index, or EUI. Central High School has 160,000 sq. ft.; so its 1990-91 EUI would be 109,820 Btu/sq. ft.

Table 5-1
BTU Conversion Factors

FUEL	UNIT	BTU PER UNIT
Electricity	kWh	3,413
Natural Gas	mcf	1,030,000
Distillate Oil (No. 2 and diesel)	gallon	138,690
Residual Oil (Nos. 4, 5, and 6)	gallon	149,690

The Btu/sq.ft. formula represents consumption over a period of time. Unless otherwise specified, the formula generally implies a period of one year.

In comparing consumption for this March compared to the previous March, you may need to divide the consumption into Btu/sq.ft./day since meter readings seldom cover the same number of days.

Gathering Building Data with Central Meters

Attempting to gather consumption data for a particular building can be a problem if the building is not individually metered. Perhaps the simplest way to gather data on a single building, which is tied into a central system, is to have the plumbers rig a "traveling meter" to check fuel consumption. Similarly, electrical consumption can be measured with an amprobe and a strip recorder. Neither item is very expensive nor difficult to use.

Degree Days

Under some conditions, it is advisable to qualify the Btu/sq.ft. measurement to allow for the impact of outside temperatures on energy consumption. If, for example, you modified your building to save heating fuel just before a severe winter, raw data on fuel consumption might not provide a true picture of what has been accomplished.

The common practice is to refer to degree days (DD), which are calculated by determining the difference between an established indoor reference temperature and the mean daily temperature. The mean daily temperature is obtained by averaging all the outside temperature readings taken over a 24 hour period. Degree day information for a given locality can be obtained from the National Oceanic and Atmospheric Administration (NOAA), commonly referred to as the national weather service. Other sources of degree day data are the local utility or a nearby airport.

In order to use degree days in relation to Btu/square foot, the total degree days must be for the same period of time as the recorded fuel use. The Btu/sq.ft./DD can be determined by taking the Btu/sq.ft. (found by using the above formula) and dividing it by total degree days for the specified time period.

The use of degree days is an optional procedure. For a given facility, using degree days may be inappropriate, or not worth the trouble, for several reasons. First, degree days are a measure of the difference between the average temperature over 24 hours and a fixed reference temperature, usually the 65° F. NOAA uses. The building may not be used 24 hours a day and the desired inside temperature is seldom 65° F. Varying inside temperatures and shutting off equipment at night and on weekends require that the DD value be adjusted, or qualified. Otherwise, the figures can be misleading.

It should also be noted that total reliance on degree days for weather adjustment also ignores humidity and/or wind, which in some circumstances, can be even more significant than the temperature. The inclusion of degree days should not be taken to mean that weather conditions have been fully treated.

In areas where temperatures vary widely in a 24 hour period to the extent that heating is required in the morning and cooling in the afternoon, the *mean* daily temperature will average out the extremes and will not reflect the loads they place on the heating ventilating or air conditioning (HVAC) system.

Finally, under some conditions, outside temperature variations may have such a small proportional effect on consumption that you will find it is not worth getting the data and factoring it into the Btu equation. For example, in a building with good thermal resistance and a high internal load, such as a school, the outside temperature variations may have a relatively small effect.

Degree days can prove very useful in comparing consumption levels for the *same building* from one period to another. As mentioned before, energy conservation benefits may not be revealed in the calculations unless the data are weather adjusted. Since the building is being compared to itself, most of the variables listed above are constant and, therefore, can be disregarded in single building comparisons.

Worksheet 5-1 is designed to check your understanding of energy consumption calculations. Answers are provided in the References section for calculations using 1,417,000 kWh, 930 mcf, and 47,000 square feet. To make the exercise most effective, supply consumption information from your own facilities and work through the process as well.

Buying BTU

In the discussion of conversion factors, it became apparent that not all units of energy deliver the same number of Btu. It is equally important to recognize that all Btu do not cost the same amount. Table 5-2 represents the national average delivered price per million Btu (MMBtu) by fuel type published in 1990 by the U.S. DOE and the Department's projected costs for the years ahead.

Local conditions will vary from the national data presented in Table 5-2. To get a better read on local costs per MMBtu, you should check with the state energy office or ask the representative(s) of your electrical and fuel suppliers. Electrical costs per MMBtu should include demand charges as well.

Since reducing energy *costs* is the ultimate objective, managers

Worksheet 5-1
Calculating Energy Consumption

Building ________________ Period __________ to __________

(month or year)

Gross square feet ________

UNITS CONSUMED	BTU/UNIT		TOTAL BTU
Electricity	kWh ×	3,413	= ________
Natural gas	mcf ×	1,030,000	= ________
Distillate oil	gal ×	138,690	= ________
Residual oil	gal ×	149,690	= ________
Other	×	________	
		TOTAL	________ BTU
Divided by ________	sq. ft.		= ________ BTU/SQ.FT.

should recognize that saving electrical Btu saves more money than saving Btu generated by gas, propane, or oil. Energy costs per square foot, or an Energy Cost Index (ECI), should be a major factor in considering which energy conserving options to undertake. Similarly, the cost of Btu should be a major determinant in any fuel conversion deliberations, or in determining heating sources in new construction.

Special rates may be offered by an energy supplier to persuade a building owner to convert to a given fuel or to encourage its use in new construction. Your first step should be a discussion with your current supplier. You may be in for a pleasant surprise, it is not uncommon for the current schedule to be adjusted so conversion to another fuel source is no longer as attractive.

Similarly, if an engineer recommends fuel conversion from electricity to natural gas for heating, it is important to check the engineer's calculations to be sure the new costs are figured on a rate schedule other

Table 5-2
Fuel Price Summary: Commercial Sector
(1990 Dollars per Million Btu)

Year	Distil-late	Residual Fuel Oil	Other Petroleum	Natural Gas	Elec-tricity
History					
1970	3.00	1.21	4.59	2.04	16.72
1980	8.60	5.49	10.42	4.42	21.61
1990	5.65	3.15	9.17	4.77	22.49
Projected					
1995	5.33	4.07	9.54	5.20	22.65
2000	5.69	4.32	10.06	5.68	21.55
2005	6.69	5.13	11.37	7.12	21.97
2010	7.29	5.56	12.14	8.04	22.44

* *Annual Energy Outlook*, 1991, Table A3, page 46-47, U.S. Department of Energy reference case forecast. Historical prices through 1980; 1990 oil prices preliminary estimate; other 1990 prices derived for DOE Energy Information Administration (EIA) data; other prices are EIA projections. Annual sector growth 1989-2010 estimated at 1.2% per year.

than the all electric schedule presently in use. Unless changes in rate schedules are taken into account, the conversion calculations will suggest greater potential savings than will be realized.

In comparing costs of energy by MMBtu, another factor you will need to consider in your calculations is the efficiency of the fuel in delivering heat or work. Electricity comes through the wall ready to go to work; its efficiency is 100 percent. Delivered natural gas or oil must use a boiler to convert the fuel to usable heat. Efficiency for natural gas, for example, is more apt to be about 70 percent. Fuel efficiencies can be obtained from your state energy office.

Of lesser importance is the initial cost of the heating plant and its operations and maintenance needs over time; nevertheless, and estimate of these costs should be factored into the comparison calculations. Worksheet 5-2 is presented below so you can compare the cost of Btu for various fuels with the efficiency factor incorporated in the calculations.

The figures used in Worksheet 5-2 are from recent billings for a private high school in New York. You are encouraged to go through a similar exercise using your own utility data, particularly if you need to prioritize several measures or you are considering a fuel conversion measure.

Whenever energy decisions are made, the cost of the energy saved must be considered. Going beyond the units of fuel saved to the energy value, or the cost/Btu of that fuel, provides the best basis for such decisions.

WORKSHEET 5-2

FINDING THE EFFECTIVE COST OF ENERGY

ELECTRICITY:

586,688 KWH X .003413 X 1.0 = 2,002.37 MMBTU
(from utility bill)

$ 45,755 / 2,002.37 MMBTU = $ 22.85 / MMBTU
[a]
(from utility bill)

~~~~~~

OTHER FUEL; NATURAL GAS:

9,538 MCF X 1.03 X .7* = 6,876.89 MMBTU
(from gas bill)

$ 45,904 / 6,876.89 MMBTU = $ 6.67 / MMBTU [b]
(from gas bill)

~~~~~~

COMPARE COST PER MMBTU:

ELECTRICITY $ 22.85 / MMBTU [a]

NATURAL GAS $ 6.67 / MMBTU [b]

* In projecting the effectiveness of fuels, engineers typically calculate from known cost/Btu. They, therefore, divide by the effectiveness factor instead of multiplying as shown in Worksheet 5-2.

CHAPTER 6

OPTIMIZING INDOOR AIR QUALITY AND ENERGY EFFICIENCY

Indoor air quality has become a major concern for building owners and facility managers. Indoor air problems can increase absenteeism and reduce productivity. Small losses in productivity can outweigh fairly large gains in energy efficiency; so when in doubt, owners and operators favor measures that can maintain or enhance indoor air quality.

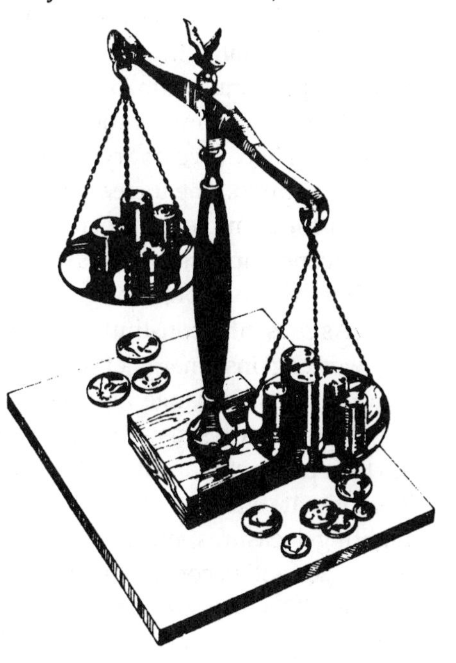

In addition to these economic concerns, there are serious legal issues *and* no liability insurance protection to cover pollution lawsuits. Finally, there is the simple fact that humane concerns outweigh other factors. *When weighed against energy efficiency indoor air quality will take precedence, every time.*

Concern over indoor air quality presents two serious implications for performance contracting. First, concerns about indoor air quality will heighten resistance to energy efficiency measures. Most articles on indoor air quality lay the blame on the doorstep of the energy efficient building. Energy efficiency, however, is not the antithesis of indoor air quality. In response to the higher energy costs of the 1970s, renovation and new construction brought tighter envelopes and increasing reliance on mechanical ventilation. By reducing the intake of outside air, existing pollutants have been concentrated and their effects on humans have become more obvious. The contaminants were already there: the energy efficient building has merely served as the messenger. Furthermore, outside air can be introduced through mechanical systems.

But the seeds have been planted. For the organization seeking performance contracting, the energy leadership will be harder to maintain. The rank and file with allegiance to the status quo will come armed with their clippings, and the nay sayers will have "grave concerns."

The simplistic argument that the energy-efficient building is to blame for our indoor air quality woes has led to simplistic, misguided solutions. In particular, "energy culprit" thinking has exaggerated ventilation as the cause and the cure.

Ventilation as a mitigating strategy usually has a negative impact on energy efficiency. The revised ASHRAE standard 62-1989, "Ventilation for Acceptable Indoor Air Quality," increases outside air from 5 cubic feet per minute (cfm) per occupant to 15 cfm/occupant in classrooms and 20 cfm/occupant in offices. Computer simulations of ASHRAE 62-1989 compliance reveal varying levels of increased energy consumption will be required. The increase in energy consumption for our nation's schools, for example, has been put at 20 percent. Others have projected energy consumption increases up to 50 percent depending upon climate, building construction, configuration and function.

As presently constituted, indoor air quality (IAQ) remedies usually work against energy efficiency and performance contracting. IAQ remedies could change any heating, ventilation or air conditioning (HVAC) measure's cost-effectiveness significantly. The payback time is apt to be protracted and it will take longer to recover energy investments. The ultimate effect will be shallower HVAC measures and less savings. Lighting measures and other measures not impacted by greater air exchanges will become more attractive in comparison.

ESCOs will also find more sales resistance to energy efficiency and, therefore, performance contracting in areas where there has been extensive press coverage of an indoor air problem.

The following ESCO "shoulds" provide marketing considerations for ESCOs and suggest expectations for potential customers:

- Energy service companies *should* offer ways to offset increased consumption due to IAQ treatments and/or provide technology that meets the ASHRAE 62-1989 standards under lower ventilation conditions.

- ESCOs *should* become conversant in IAQ concerns and remedies and their real relationship to ventilation and energy efficiency. The ESCO who can offer the customer facility audits that embrace energy efficiency *and* IAQ will have the competitive advantage.

For the organizations seeking ESCO services, IAQ credentials and assurances should be part of the performance contracting solicitation and evaluation process. Having the IAQ bases covered is good business *and* good legal protection.

ESCOs also need to be somewhat conversant in the real relationship between IAQ and ventilation. The most common cause of IAQ problems is poor maintenance in the building's HVAC system. In fact, the relationship between energy and IAQ is more apt to rest on the typical cuts in maintenance to pay the utility bill. The advent of more sophisticated energy equipment and the conspicuous absence of appropriate training adds to the problem. It helps, therefore, to recognize that mechanical system selection can effect indoor air quality both directly and indirectly. The type of mechanical system, its components and its accessibility can contribute to poor maintenance.

IAQ AND ENERGY EFFICIENCY

Ventilation can be beneficial when the contaminant or its source can't be determined, as an intermediate step until action can be taken, or when source mitigation strategies would be very costly. Specific applications of ventilation; e.g., localized source control or sub-slab ventilation to control radon, are valuable control measures.

Many would put IAQ and energy efficiency in an adversarial relationship. They are not. They are part and parcel of the same goal for designers, builders, owners, facility managers and operators: providing a comfortable, productive work place as cost-effectively as possible.

There are measures that can be taken to improve IAQ that are energy neutral and some measures that improve IAQ *and* energy efficiency. A sampling of these measures will help the practitioner spot others.

Energy Neutral IAQ Remedies

Many of the ways to improve the quality of indoor air focus on maintenance. The growing cost of poor quality indoor air gives added importance to preventive maintenance (PM). In fact, today PM carries increasing overtones of another PM: preventive medicine, for preventive maintenance can avoid health problems. Some PM measures representative of low cost source control and training are:

- checking duct linings, which are breeding grounds for microbials, especially in areas of high humidity;
- exercising care that steam for humidification has no contact with boiler additives;
- removing partitions impeding critical air flow; or using through-the-wall fans where applicable;
- checking condensate pans in unit ventilators, fertile sources for biological growth of fungi and bacterial organisms; treating with algicide as directed on container;
- exercising great caution in the use and storage of toxic chemicals in operations, maintenance, pest control, kitchens, etc. and using less toxic chemicals wherever possible;
- checking common reservoirs for microbial contamination; i.e., flush toilets, humidifiers and ice machines; or
- positioning air intake grills so that the re-entry of fumes from the building's own exhausts or other avoidable contaminants, such as fumes from delivery alleys and loading docks, are not a factor.

This list of measures for IAQ control argues well for an effective preventive maintenance program.

Measures That Enhance IAQ and Energy Efficiency

Many factors that have a negative impact on energy efficiency also have an adverse effect on IAQ. Correcting such situations can improve energy usage and enhance the quality of the air. A few examples will illustrate the point.

- Poor maintenance of pulleys, belts, bearings, dampers, and other parts of the mechanical systems can increase resistance causing a decrease in air supply. Good maintenance improves both energy efficiency and IAQ.

- Water damaged insulation, ceiling tiles, rugs and internal walls support biological growth. Wet materials nullify insulation properties. Replacement removes breeding grounds and increases thermal resistance.

- The routine replacement of filter media facilitates the filtering of particles and avoids the resistance to air flow caused by clogged filters, which left unattended reduce energy efficiency.

- Humidity control can reduce the likelihood of mold while contributing to comfort and energy efficiency.

- Frequent causes of combustion contaminants are defective central heating systems in which the exhaust is not vented properly, or there are cracks or leaks in equipment. Defective or poorly maintained systems are less efficient and, at the same time, polluters.

IAQ Conscious Energy Efficient Design

'Engineering out' materials and conditions known to contribute to indoor air pollution in new construction or renovation work may prove to be the most "energy efficient" measure of all.

Siting of a facility should consider potentially negative influences on IAQ, such as the quality of the soil, particularly in radon affected areas or, the proximity of roadways and vegetation which may, depending on the type and placement, filter out contaminants or serve as a source of allergens.

Some building materials may produce varying amounts of air contaminants. The careful selection of construction materials and furnishings can help alleviate indoor air problems. Among the things to watch for are formaldehyde emission from pressed wood products and other volatile organic compounds (VOC). Standards can be specified that meet Federal regulations (CFR 24, Part 3280) or industry association guidelines (NPA 8-86 and HPMA FE-86). Caulks, sealants and adhesives have various levels of VOC emissions depending on the compound. Building, pipe and duct insulation containing urea-formaldehyde can have an adverse effect on IAQ. Paints have highly variable mixtures of VOCs and have high release rates during the short-term curing process. Paint should be applied and cured in well-ventilated conditions.

Care should be taken during construction to insure that IAQ is not compromised before the building is used. Materials and furniture should have had sufficient "curing" before the building is occupied. This means timing construction so commissioning can be completed before the building is brought into use. This will reduce off-gassing and other noxious effects of new materials or equipment during occupancy.

In addition to the consideration given to the thermal environment and humidity, air flow patterns need careful HVAC designer attention. Ventilation effectiveness, the ratio of outside air reaching occupants compared to total outside air supplied to the space, determines the capability of the supply air to limit the concentration of the contaminants.

Filter replacement access is an important consideration. Filters used should be rated for efficiency; i.e., tested in accordance with ASHRAE Standard 52-76. Higher efficiency filters (above 30 percent) can be effective at reducing levels of respirable particles. Filters can also be specified based upon their removal efficiency for particles of a specific particle size. Filter location and media selection is best left to specialists.

Operating the building and its equipment to meet environmental needs requires good maintenance and conscientious custodial practices carried out according to manufacturers' schedules and directions.

Codes and standards are useful references in new construction and may be mandated by state or local laws. In view of the growing concern regarding IAQ, these are apt to be revised and additional regulations promulgated. ESCOs and administrators need to take the initiative in ascertaining the current status of IAQ regulations.

Keeping An Eye on Legal Implications

In today's litigious society, the specter of legal action always looms large. Sick buildings clearly offer fertile new territory. Attorney/engineer Gardner has predicted, "Indoor Air Quality will be the toxic tort of the '90s."

Way back in 1988, the *Wall Street Journal* reported:

> "SICK BUILDINGS" LEAVE BUILDERS AND OTHERS FACING A WAVE OF LAWSUITS
>
> More office workers are filing lawsuits, claiming they were made ill by indoor air pollution from such things as insect sprays, cigarette smoke, industrial cleaners, and fumes from new carpeting, furniture, draperies and copiers.

Concerns have continued to mount as indoor air quality issues haunt design professionals, builders, owners and managers. *The Professional Liability Perspective,* published by a prominent California insurance broker, warned design professionals,

> Indoor pollution is seen by many close to the profession as something of a new frontier for the underemployed at the plaintiff's bar. Your challenge is to take steps to remove yourself from the path of the almost inevitable stampede.
>
> The potential liability problems posed by indoor pollution are compounded by the fact that the pollution exclusion in your policy of professional liability insurance is all-compassing. It extends to every form of environmental contamination imaginable. <u>The risk is simply not insurable</u>. [underscore supplied]

Where does the liability reside? The insurance broker's newsletter indicated that the problem rests primarily with the owner. To get out of the "path of the almost inevitable stampede," architects and engineers were urged to recommend to owners that they retain qualified air quality consultants.

The advent of such newsletters as the *Indoor Air Pollutant Law Report* provides further evidence of litigious clouds on the horizon for design professionals, builders, owners and facility managers.

In an effort to improve energy efficiency, you need to be sure neither you, nor your ESCO, puts you in the "path of the almost inevitable stampede." The following suggestions are offered to help avoid some of the legal risks associated with indoor air problems, and to reduce owner/manager vulnerability. The suggestions are not designed to supplant those of legal counsel. However, should legal action be forthcoming, they offer ways to make life a little easier for your organization as well as your engineers and attorneys.

Ignorance Is No Excuse

Before any attention is paid to specific indoor air quality (IAQ) concerns, it is well to remember that the court records are full of cases where the defendant had the information and failed to act, or the defendant *"should have known."* This does not mean that you or any of your staff should necessarily don the mantle of an industrial hygienist or an environmental engineer. It does mean that your organization should be able to demonstrate that it took advantage of basic information readily available, such as *Managing Indoor Air Quality* for owners, managers and those who offer them technical support.

IAQ concerns have erupted around us, the means to identify, control and treat a myriad of pollutants has mushroomed. The IAQ field is no longer new, and experience in almost every type of facility exists to help us. Still, large gaps remain in our IAQ knowledge and new information is emerging rapidly. Newsletters, such as *Indoor Air Quality Bulletin,* offer a way to keep on top of what is happening. Through publications, like the *Indoor Air Pollutant Law Report,* owners and managers can be privy to the areas of potential litigation being eyed by attorneys.

Knowing is not enough. Effective IAQ management strategies require a demonstrable effort to implement applicable procedures. Industry has been dealing with indoor pollutant problems for some time. Its efforts have been primarily directed toward specific contaminants known to be part of the product or process. While commercial buildings, offices and institutions are apt to have multiple-contaminant concerns, many of the management practices honed over the years in industry are applicable to broader IAQ concerns and their legal ramifications.

Management strategies related to IAQ are centered on certain facets of communications and Sick Building Syndrome (SBS) prevention. Communication strategies, which are discussed first, are well known to effective managers and need only be adapted to IAQ-related procedures.

Complaint Response

Each IAQ episode exacts a price, human and economic. That price warrants establishing an effective complaint response procedure. A complaint that is lightly dismissed creates an emotional climate that is hard to overcome. A seemingly innocuous complaint that does not receive careful attention can grow into job dissatisfaction and lost productivity. Almost overnight, it can spread to become a disgruntled work force, a union dispute or confrontation in court.

A complaint may actually be prompted by a cold or the flu. Or, it can start with a newspaper article, coupled with someone's active imagination and encouragement from the coffee lounge brigade. On the other hand, the symptoms may be from a very real building-associated illness. In either case, it can't be ignored.

Owner/tenant relationships can suffer the same consequence and the economic impact is apt to be more immediate and decisive. The tenant moves out. Word that the building is "sick" and the management was unresponsive spreads rapidly.

In too many instances, a "brushed off" complaint can lead to a very large public relations or legal problem even though the actual health involvement may have been minimal. Unless you want to read about your problems in the newspaper, establishing procedures to respond to complaints is critical. A response tracking procedure and follow-up procedures are important management tools.

The way complaints are received and responded to creates a

climate that can offer cooperative resolution, or it can create an adversarial situation ripe for litigation.

When good IAQ management procedures are ignored, the "knee jerk" reaction is to turn up the fan, which runs up energy costs. Increased ventilation may not solve the problem, but it will most assuredly cause the ESCO headaches.

Preventive Actions

There is an old line: "If you hear hoof beats behind you, it's more apt to be a horse than a zebra." Or, to put it another way, we need to draw on our experience and not overlook the obvious. The "obvious" in this case is managing legal risks by avoiding "sick buildings" in the first place. Even though IAQ episodes have been found to be multifactorial, reviewing the records of the major IAQ investigating teams shows a commonality of problems that can serve as a sound basis for keeping a building healthy.

A wealth of information on effective preventive strategies is reflected in the table on the following page. The terminology varies among the investigators, but the problems found are quite consistent. All investigators found many problems traceable to operations and maintenance deficiencies. For instance, Honeywell used "Operations and Maintenance," Healthy Building Institute (HBI) cited "Contaminated Systems" and poor maintenance is a major contributor to the National Institute of Occupational Safety and Health (NIOSH) category "Inadequate ventilation."

Problems can also be avoided by spelling out IAQ needs in purchasing and construction specifications, including work the ESCO performs. Making the effort to determine these needs in the beginning can avoid big management and sick building syndrome headaches later.

Table 6-1.
Sick Building Syndrome Problem Frequency

Or. NIOSH	HONEYWELL	HBI
Bldgs. 529	50	223
Yr. 1987	1989	1989
Inadequate ventilation (52%) Inside contamination (17 %)	Operations & Maintenance (75%) - energy mgmt. - maintenance - changed loads	Poor ventilation - no fresh air (35%) - inadequate fresh air (64%) - distribution (46 %)
Outside contamination (11%) Microbiological contamination (5%)	Design - ventilation/ distribution (75%) - filtration (65%) - accessibility/ drainage (60%)	Poor filtration - low filter efficiency (57%) - poor design (44%) - poor installation (13%)
Building fabric contamination (3%)	Contaminants (60%) - chemical - thermal - biological	Contaminated systems - excessively dirty duct work (38 %) - condensate trays (63%) - humidifiers (16%)

CHAPTER 7

RISK ASSESSMENT AND MANAGEMENT STRATEGIES

When it comes to energy financing, discussions of risk seem to precede any attention to opportunities. Many energy financing "old wives tales" are unfounded, including:

> "An [institution] will lose... control... "
>
> "The contractor is telling the school that for the next seven years, energy systems will run in such-and-such a manner."
>
> " ...there are some definite clauses as to how energy savings are going to be calculated and how the lease will be figured."

These are direct quotes from an article titled "Expert Testimony on Energy" published in an educational trade journal in the late 1980s.

All of them imply the acceptance of some uncontrolled risks on the part of an organization. They perpetuate the mysteries that seem to surround energy financing and the sense that the organization, the customer, must take a passive role.

Such assumptions are simply not true.

If we take a look at the nature of risk and then the risks associated with energy financing in particular, it becomes clear that such risks can be assessed and managed. In fact, performance contracting provides an organizational opportunity to shed some risks that would otherwise be incurred, if you were to buy and maintain the equipment yourself.

The Nature of Risk

Risk is a normal part of our daily lives. We make risk decisions constantly from the time we wake up in the morning until we drift off to sleep at night. Do we "risk" having coffee for breakfast? Do we *risk* passing the car in front of us on the way to work? Do we use the front door of our building or the back door? Do we say a cheery "Good Morning" to our associates? With some associates, that may well be a risk.

However trivial, there is some degree of risk in every action we take... or don't take. The risk assessment and risk management process is as natural as breathing. For example, some say drinking coffee is bad for us, but for many of us, the need for our morning coffee is stronger than the idea that it might be a problem. That is risk assessment... balancing risk against benefit.

The next step is risk management. We may concede that drinking coffee could be a problem, so we reduce the risk (or manage the risk) by limiting ourselves to one cup or by switching to decaf. Or, we may decide that the risk is so slight, it can be ignored and we drink four cups. That is accepting the risk. The process becomes automatic and can be traced through in the same way for most actions we take.

Risk, as a function of energy finance, requires the same general process of risk identification, assessment, management and acceptance. Risks in a financing project are examined, evaluated, and then a determination is made as to how those risks will be managed.

Risk plays a major role in energy financing decisions and in the structure of financing packages. Perceived risks, whether real or not, can cause someone in your management to resist private sector financing of energy efficiency projects. From the energy service company perspective, risk limits the availability of financing and also determines the cost of money. *Risk ,and strategies to manage that risk ,always carries a price for someone.*

The risks in performance contracting can be identified, often quantified, and they can be managed. The process involves breaking down the risks into manageable segments and determining what needs to be done about them. The choices are whether or not to accept a risk, transfer it to someone else, or attempt to minimize or reduce it. There is a cost associated with each option. Ultimately, the most costly risk of all in energy financing may be to do nothing.

Risks in Performance Contracting

Every energy efficiency project involving capital expenditure carries with it a number of readily identifiable risk factors. These factors must be considered no matter how the project is to be funded. Who carries the risk will vary with the financing approach selected. Figure 7-1 offers an analysis of some risks by financing schemes and the cost of risk shedding.

Management of risk for you and those providing the service involves decisions as to; (1) whether certain risks can be reduced or eliminated, (2) the level of risk that is acceptable, and (3) who will bear the risks. How these risks are approached and managed may determine whether or not the project moves ahead at all. An examination of major performance contracting risk components can help you identify the dimensions of certain risks and suggest strategies for controlling those risks more effectively.

Construction Cost Risk

Whenever your organization engages in a new construction project, apprehensions are apt to emerge. Will the project come in on budget, or will the costs balloon? Is there a possibility of "over design" that will run up first cost and/or unnecessarily increase energy costs? Has the contractor provided construction performance and/or payment bonds that back the firm's commitment to complete the job satisfactorily, on time and within budget?

Construction risks can be reduced by providing professional construction management and oversight. The use of an independent

	Types of Risks	
Financing Scheme	To Organization	To Investor/ ESCO/Manufacturer
Guaranteed savings	[Obligations to lessor by ESCO for compensation] **Level of risk rests on stability**	**ESCO guarantees savings**
Shared savings	[Larger portion of cost goes into debt service; less into energy benefits. Projects under-funded.]	**Uncertain level of payment** Financing more limited **ESCO caries, debt obligation**
Vendor financing	[Vendor influence on equipment selection. Limitations on broader service.]	
	[Manufacturer may not guarantee all capital projects. Or, manufacturers may require maintenance contract.]	**Vendor risk to extent guarantee is not compensated**

bold face - risk
bracketed [ ] remarks—cost of risk shedding

Figure 7-1. Risk Analysis by Financing Schemes

consultant can help assure energy efficient construction. In performance contracting, the ESCO assumes most of these risks and there is less likelihood of over design.

Operations and Maintenance Risk

Operations and maintenance (O&M) staff will inevitably reduce the operation of equipment to their level of understanding. Energy inefficient buildings and indoor air problems are often traced to operation and maintenance, because the design phase did not consider the sophistication of the O&M staff. Or, the accessibility of the equipment. Or, once systems and equipment were installed, the level of training O&M staff received (or did not receive).

The O&M responsibilities and the related risks of project performance may be assigned to, and assumed by the energy service company... if you want to pay for it.

Evaluation of federal energy grant recipients' programs has shown that four out of five dollars saved in an effective energy program can be attributed to energy efficient practices of the operations and maintenance staff. Your O&M staff can be the deciding factor in the success or failure of a project. This is important to your management and downright scary to the ESCO.

To manage the operations and maintenance risk, the components can be easily identified and controlled. Maintenance of installed equipment should be part of the agreement, and the costs of that maintenance covered under guaranteed savings. The energy service agreement should provide for staff training on at least the installed equipment and, preferably, energy efficient O&M practices on all energy consuming equipment. The staff's capabilities should be considered when determining an acceptable level of sophistication in the engineering design.

Equipment Performance Risk

Who gets the blame when that shiny new equipment you selected just doesn't perform up to specifications? Or dies the day after the warranty runs out?

In addition to the warranties from equipment vendors, other warranty coverage can be sought to cover a longer period of time. It is reassuring if you remember that the energy service company's savings guarantees rest on the performance of the equipment. Under a guaranteed savings agreement, your equipment performance risk can be transferred to the ESCO.

Energy Price Risk

If you know for sure when energy prices are going to go up and by how much, there are folks ready to pay a very high fee for your prognostications. Contracts that rely on energy price predictions are risky business.

Aside from the upward trend typically associated with the dwindling supply of a finite source, actions in other countries as well as our own impact energy prices. Political as well as economic factors add to the instability. Depending upon the regulations that result, Congressional environmental actions could bring increases in energy prices within the decade to rival those of the 1970s.

You need to be leery of vendors using price predictions to their advantage. Engineers and vendors have been known to "bend" the utility bill. All too often, salesmen divide the total electric bill by the kilowatt hours (kWh) consumed. Many rate schedules for large users provide that the more kWh used, the less you pay per unit. When kWh savings occur, it is the cheapest kWh that is saved first; not the *average* kWh cost. It pays for you to know how the rate schedule works so that these false energy cost savings claims can be spotted. At the very least, running the figures by your utility representative will help avoid such pitfalls.

When seeking bids (by RFPs or specs), the risks associated with "creative savings projections" can be avoided by requiring bidders: (1) to use current energy prices; (2) to go by specified U.S. DOE projections, or (3) by defining the escalation rate to be used.

In performance contracting the contract should also clearly define the conditions for sharing increases or drops in price. If prices escalate, who enjoys the extra savings? If prices drop, who takes the hit? Increasingly, contracts have a price floor to avoid the wholesale damage ESCOs suffered in the 1980s. As a potential customer if you expect to share in any price increase benefits, be prepared to share in the downside risk.

Establishing Base Year Data

A base year is a bench mark from which savings are calculated. It needs to be fair and equitable to all parties.

Typically, an energy service company prefers to form a base year by averaging energy consumption in a given facility for the last two or three years to smooth out irregularities in building use, weather, etc. However, nothing is locked in.

You should always decide what is in your best interest and negotiate from there. If you put in new thermal windows last year, for example, using a three year average would, in effect, pay the ESCO for some of *your* savings.

Looking to the future, you need a "re-open" provision in the contract to allow an adjustment in the base year for major changes in occupancy, function, and square footage, etc. Long dissertations in contracts about degree days are usually, as the Bard would say, "Much Ado About Nothing." The three year average generally levels out such variations. (Degree days are discussed in Chapter 3.) Exceptionally warm or cold weather can be treated in a "re-open" clause.

Changes in the building occupancy or function can also have an effect on demand costs or the rate of consumption. Utility costs associated with the *rate* at which electricity is used are typically 100-fold greater than the cost per kWh, the *quantity* you use. The quantity, kWh, might typically cost around $0.08/kWh, while the rate charge per kilowatt (kW) is frequently $11.00/kW or more. How variations in demand saving is handled should be included in any energy service agreement.

The services provided by an ESCO frequently reach beyond energy cost savings and offer the customer broad operational savings as well. To reflect this range of services, and ESCO may calculate the base year and savings in this broader context, incorporating operational savings in the calculations.

Occupancy Risks

The best energy efficient system can be rendered almost useless by building occupants, who exercise a little ingenuity. Assessing this level of risk, which can defeat the best efforts by the O&M staff, is important to you and the performance contractor. The greatest risk rests with the ESCO who has guaranteed your savings. But the customer loses, too. It is an area that graphically illustrates the critical role administrative commitment and leadership can play. It underscores the sense of partnership that must prevail in successful performance contracts.

Building Control Risks

The most feared, but easiest to manage, risk is building control. You should decide in advance what are acceptable standards for building operation; e.g. temperature parameters for heating and cooling, illumination levels for lighting, run time for equipment, etc., to meet occupant comfort, safety, and productivity needs. Then, be sure that they are clearly set forth in the agreement. Despite the observations of the gentlemen quoted at the beginning of this chapter, you need not, and should not, give away any more control than you want to. The more control the customer keeps, however, the more risk the ESCO generally has to incur. *Money follows risk*. The cost of the ESCO assuming that risk must be paid.

Engineering/Technical Risk

Is the engineering approach to the project correct? Have any major energy conserving opportunities been overlooked? Are the projected savings based on reasonable assumptions? What happens if the engineers are wrong? Traditionally, an engineer is paid a percentage of the total cost of the job and the engineers are not held accountable for their savings projections. In performance contracting, most of the technical risk is assumed by the energy service company. There is some comfort in knowing the ESCO has *their* money riding on the engineer 's decisions. The ESCO 's engineer is held accountable. If he or she wants to stay employed, the estimates need to come pretty close to the mark.

With all of the risk shedding that performance contracting can offer, your organization's greatest risk usually rests in the selection of the energy service company.

Energy Service Company Selection Risk

Nearly all of the risks you might typically incur in a search for energy efficiency can be assumed by the energy service company. The selection of an energy service company, however, constitutes a risk in, and of, itself.

The best way to manage this risk is to first determine what your organization needs, and then establish the criteria for selecting an ESCO that can best meet those needs. The level of risk is dependent upon the contractor's qualifications to do the job; i.e., technological capability, personnel qualifications, financial stability, and overall track record. The simple process of getting references, checking them, and just calling colleagues is still valid. The process is markedly similar to hiring qualified professional staff. (See Chapter 10 for information on evaluating potential performance contracting partners.)

A successful energy efficiency project requires that all these risks, and more, be identified, evaluated and managed. The stakes, in many instances, go beyond the risks. Public administrators, for example, exist in the "fishbowl" of public scrutiny. For them, it can be a very real political risk should anything go seriously wrong. Often, the unquantifiable political risks pose a more lasting consequence than those with direct dollar impact.

"Do Nuthin'" Risk

With risks of the unknown on the horizon, it would seem prudent to sit tight, do nothing and just keep paying the energy bill. Not so! There are very real risks... and costs... that go with "doing nuthin'" about energy efficiency.

The risks are both financial and political. The Cost of Delay can be readily calculated as discussed in Chapter 3, "The Bottom Line." What you have paid for wasted energy that could have been saved compounds itself over time, and as the months go by, the numbers just get bigger and bigger. That's the money side. But suppose that someone becomes aware of the fact that you have been needlessly spending valuable, and scarce, dollars on wasted energy: money that could have been used for educating students, serving patients, providing taxpayer relief, increasing dividends or giving the employees a raise—all because you chose to do nothing to stop the high cost of energy waste. That's a big political risk! And a large personal risk; because it could mean your *job*!

Managing Risks Through Performance Contracting

There is less risk through performance contracting than if you had attempted to purchase and maintain the equipment in-house. In addition, the ESCO takes most of the energy worrying off your hands and provides the money to get the job done. It sounds like it is too good to be true. Not quite. There are risks. There can be problems. But if approached realistically in a business-like manner by both parties, a contract which requires a certain level of performance; i.e., performance contracting, can be an extremely valuable tool.

In the very simplest of terms, utilizing performance contracting means deciding what you want to get done, then finding someone who will do the job, investing *their* money, with payment based upon the success of the project. The procedures need not be complicated. Performance contracting has proved to be very effective for a growing number of organizations.

In the final analysis, if the manufactured layers of complexity and all the legal mumbo jumbo are peeled away, an energy service agreement is little more than an installment/purchase paid for out of guaranteed future energy savings.

The key to successful performance contracting is assessing risks and weighing the costs of those risks; then, accepting, assigning, reducing, or avoiding them; i.e., risk management. But then, that's what we do every time we pick up a coffee cup.

CHAPTER 8

ASSESSING NEEDS AND SETTING PROJECT CRITERIA

With nearly two decades of experience in energy conservation and financial planning, there have emerged common elements to guide today's planners. While the framework for self-assessment is rather generic in nature, precise responses based on local conditions can provide the basis for a plan that meets an organization's specific needs.

On the following pages, *The User's Manual to Energy Cost Savings*, is designed to serve as a self-assessment guide that will enable you to identify and consider those elements that have proved to be basic to energy financial planning. This "User's Manual" to energy financing is not intended to be all inclusive. When you sit down with your colleagues, it will give you a place to start. Questions raised in this section and throughout the book will suggest other items that may need to be considered to meet specific needs.

Finally, needs change over time. Neither the responses to the questions posed or the plan that emerges should be considered static. Additional investigation of certain energy financing concerns may also indicate that some initial responses need further consideration.

Following *The User's Manual to Energy Cost Savings* is a discussion of the financial planning that can emerge from this assessment process. A plan outline is offered to suggest the form such a plan could take. The chapter concludes with a section on setting criteria for energy action and solicitation procedures which should logically grow out of the self-assessment and planning process.

THE USER'S MANUAL TO ENERGY COST SAVINGS: A SELF-ASSESSMENT GUIDE

I. Can we make money through energy management?

NEED A ROUGH IDEA OF YOUR ENERGY ORGANIZATION'S SAVINGS POTENTIAL?

[Typically those who have done very little, can usually "make" 25 percent of their utility bill.]

Calculate:
Last year's energy costs $__________ × .25 = $ ________________

WANT SOMETHING MORE EXACT?

Check existing energy audits for cost saving potential and extrapolate the savings potential to similar buildings.

Projected savings for recommended measures $ _______ divided by buildings sq.ft. = $ _______________ savings/sq.ft.
$ __________ savings/sq.ft. × __________ total/sq.ft. of all facilities = $ _______________ total savings potential.

Or,
Compare utility costs/sq.ft. in your facilities to average costs/sq.ft. for similar buildings in region. Your state energy office may be able to supply typical cost data for similar buildings. An ESCO in the area might also share with you the cost/sq.ft. that is required to be candidate for performance contracting.

Calculate:
Our utility costs/square foot [sample buildings] $ __________ (a)
Average cost/sq.ft. __________ (b)
Difference [(a) minus (b)] __________ (c)
If a sample was used:
Multiply total sq.ft. ___________ × _________ (c) = $ _________

THE USER'S MANUAL TO ENERGY COST SAVINGS: A SELF-ASSESSMENT GUIDE

II. Energy Efficiency?!? Who cares!

LIST KEY FINANCE AND ENERGY DECISION-MAKERS IN YOUR ORGANIZATION (INCLUDE INFLUENTIAL BOARD MEMBERS IF APPROPRIATE):

1. ______________________________
2. ______________________________
3. ______________________________
4. ______________________________
5. ______________________________

WHAT DO THEY WANT THAT COSTS MONEY?

(An MRI? Higher salaries? Sales training? More computers? Greater dividends?..?..?)

WHO .... WANTS .... WHAT

1. ____________ .... ____________
2. ____________ .... ____________
3. ____________ .... ____________
4. ____________ .... ____________
5. ____________ .... ____________

[Public institutions; i.e., schools, local governments, should remember to include items with strong public appeal as well.]

COULD ENERGY COST SAVINGS "PAY THE FREIGHT"?

[Compare projected savings calculated in "I" on the preceding page to the cost of these special wants; i.e., divide potential savings by the cost of the computers, or ____________ .]

WHAT KIND OF COMMUNICATIONS STRATEGIES DO WE NEED TO BUILD THE SUPPORT, COMMITMENT?

[Hint: Talk money; not energy efficiency or Btu saved. Even better talk their needs; not money.]

THE USER'S MANUAL TO ENERGY COST SAVINGS: A SELF-ASSESSMENT GUIDE

III. How much are we losing each year by doing nothing?

WHAT IS IT COSTING US TO POSTPONE . . . . ? ? ?

Calculate:

ENTER potential energy savings (from I)	$	________
ADD estimated operational savings (including maintenance, parts)	+	________
TOTAL PROJECTED SAVINGS	=	________
SUBTRACT annualized cost of energy saving equipment and installation (Total cost/years payment will be made)	–	________
COST OF DELAY EACH YEAR!	= $	________

HOW MANY UNIFORMS, TEXTBOOKS, ETC. COULD WE BUY WITH THE MONEY WE ARE SPENDING ON WASTED ENERGY?

Calculate:

Cost of Delay/cost of EACH uniform = ________

or, COD/average teacher salary = ________

[Hint: Check the "who wants what" on the preceding page, get the prices for the items; then divide Cost of Delay by these prices.]

When the answer is, "It's a good idea, but let's wait until we can do it ourselves," check your organization's track record for idea to installation timing.

Calculate the Cost of Delay for the typical "idea to installation" in your organization. Compare the COD to the service fee amount that would go to an ESCO over a typical contract life (7 - 10 years). Consider time value of money in making comparisons.

THE USER'S MANUAL TO ENERGY COST SAVINGS: A SELF-ASSESSMENT GUIDE

IV. What can we do to save energy dollars?

CHECK PROJECTS SUGGESTED FROM WALK-THROUGH BUILDING SURVEYS:

Based on surveys, what can you do right away?

[ ] energy work by operations and maintenance staff
[ ] low cost opportunities

[Remember: There is an inclination to do the low-cost items before seeking financing; however, if the organization waits too long to do its O&M work before seeking private financing, the Cost of Delay will be greater than the O&M generated savings.]

COMPREHENSIVE ENERGY AUDITS:

Do the surveys suggest that professional energy audits are warranted?

Do other buildings need comprehensive energy audits?

LIST BUILDINGS NEEDING ENERGY AUDITS:

HOW SHOULD WE PAY FOR ENERGY AUDITS?

[ ] engineering consultant [you pay 100%]

[ ] using a federal energy grant (for schools, colleges and hospitals) [you pay 50%; if you qualify as"hardship," you pay 10-50%]

[ ] have an ESCO do it [no up front cost] as part of its contracted services. You may need to pay a nominal fee to have an independent energy engineer evaluate the ESCO's audit quality. These costs can be assigned to project costs.

THE USER'S MANUAL TO ENERGY COST SAVINGS: A SELF-ASSESSMENT GUIDE

V. What makes sense?

HOW DO WE MEASURE WHAT'S COST-EFFECTIVE?

$$\text{Simple Payback} = \frac{\text{Initial investment}}{\text{Projected Annual Savings}} = \_\_\_\_\_\_ \text{ years}$$

OTHER CONSIDERATIONS (check applicable item) for further examination):

[ ] Are operations and maintenance costs a factor?

[ ] Is equipment getting old? Does it need major work or replacement? Would automated controlling of existing equipment present problems?

[ ] Are there other needs that have energy implications; i.e., adding insulation to needed re-roofing?

[ ] Does the life of the building warrant the investment?

WHAT SHOULD YOUR RETURN ON INVESTMENT BE?

[ ] What is the cost of independent financing? __________ %

[ ] What are your present investments yielding? __________ %

[ ] Does your organization have endowments, reserves or other resources that could be invested in your energy efficiency for a greater return? ______________________________

If yes, what is the yield? ______________________________ %

[ ] What rate of return do you need to make using <u>your</u> fund's good business? ________________

[ ] What is the opportunity value of the money? ____________

____

FOR PUBLIC INSTITUTIONS:

[ ] What would be the effect on your level of indebtedness?

____

[ ] Should you use tax exempt financing so more of the invested amount will go for energy saving modifications?

IF IT MAKES SENSE FINANCIALLY, DO YOU HAVE WHAT IT TAKES TO GO IT ALONE?

[ ] Energy expertise: "in-house" engineering

[ ] Energy audits in hand

[ ] Manpower—administration, maintenance, etc.

[ ] Organizational momentum to keep on top of energy monitoring and maintenance needs.

THE USER'S MANUAL TO ENERGY COST SAVINGS: A SELF-ASSESSMENT GUIDE

VI. What risks are you willing to take to make money?

WHAT ARE THE MAJOR RISKS YOU FACE IN ENERGY FINANCING? [See Chapter 7, "Risk Assessment and Management Strategies."]

[ ] Engineer's predictive ability

[ ] Equipment performance

[ ] Energy price changes
(determines level of cost savings over time)

[ ] ________________________________

HOW MUCH ARE WE WILLING TO "PAY" FOR A PRIVATE FIRM TO ASSUME SOME OF THESE RISKS?

WHICH ONES?

WHAT IS THE MANAGERIAL/POLITICAL RISK OF CONTINUING TO SPEND PRECIOUS OPERATIONAL DOLLARS OR TAX DOLLARS ON WASTED ENERGY?

THE USER'S MANUAL TO ENERGY COST SAVINGS: A SELF-ASSESSMENT GUIDE

VII. Performance Contracting?

WHAT ARE THE ADVANTAGES IN USING PRIVATE SECTOR FINANCING?

[ ] immediate positive cash flow
[ ] no need to increase indebtedness
[ ] opportunity value of your money
[ ] you get their energy expertise

[ ] "they" worry about energy and leave you with time for other things

[ ] you get needed capital improvements **NOW** from future energy savings

[ ] unit costs will go down, contributing to our competitive advantage

WHAT ARE THE CONCERNS, BARRIERS?

[ ] what limitations or management concerns exist?
[ ] physical plant concerns?
[ ] technical problems?
[ ] financial concerns?
[ ] procedural limitations?

NOT SURE?

CONTACT: Your state energy office, The National Association of Energy Service Companies, or an ESCO (See list in Appendix B.)

Ask about other performance contracts in your area.

THE USER'S MANUAL TO ENERGY COST SAVINGS: A SELF-ASSESSMENT GUIDE

VIII. Doing your own thing?

HOW DO YOU GET JUST WHAT YOU NEED?

FIRST, YOU NEED TO FIGURE OUT EXACTLY WHAT YOU NEED.. THEN ASK FOR IT!

SERVICE OPTIONS TO CONSIDER:

[ ] equipment ownership

[ ] lease/option to buy

[ ] maintenance service - installed, existing equipment

[ ] O&M training

[ ] energy accounting software

[ ] preventive maintenance software

[ ] minimal risk

[ ] staff/public information support

[ ] energy-related work (which is not as cost-effective); e.g., boiler replacement, roof replacement

[ ] non energy-related work

CONTRACT TERMS TO CONSIDER
[See Chapter 12, "Contracts"]

[ ] comfort parameters (light levels, temperature ranges)

[ ] length of average payback

[ ] formulas; treatment of base year and variables such as weather, occupancy

[ ] performance, savings guarantees

[ ] share of savings

[ ] length of contract (contract length is influenced by paybacks, services requested, & your share of savings)

Energy Financing Plan

Once self-assessment procedures are completed, you may wish to crystalize the findings into some type of energy financing plan. This plan may address broad purposes only, or it may become quite detailed. It should at least state the basic financing procedures that have been accepted; e.g., the use of traditional funding sources for specific measures and/or the determination to seek private sector funds for energy efficient capital improvements. The plan might also address concerns/risks and the levels acceptable to the organization. A discussion of other items such as services required, maintenance needed or payback parameters, may be desirable. All of these items can be discussed within the context of the various financing options available.

Who will see/use the plan? The expected use of the plan will also serve as a major determinant in its format and its contents. If it is to be used as a management tool, then specific operational concerns may be more appropriate than they would be if the document were destined for board approval at a public meeting.

An outline of an energy cost savings plan is presented below to help guide the development of an energy financing strategy. If you don't have the financial expertise in-house to weigh the options, your accountant or banker can be of help to you.

Energy Costs Savings:
An Outline for Financial Plan Development

I. Statement regarding the current level of energy waste and the potential economic benefit to the firm, institution or other organization.

II. Criteria for financing energy reduction projects:

  A. Cost-effectiveness

    1. Project can be financed out of the avoided utility costs; i.e., energy savings will be greater than the debt service payments; and

    2. Project meets return on investment (ROI) hurdle rate.

  B. Establish a target rate of return for energy efficiency investments.

III. State risks associated with options:

A. Identify the risks typically associated with each energy financing approach and such concerns as:

1. future energy price changes;
2. technical concerns, including engineering risks, uncertainties related to equipment performance, and operations problems;
3. private firm's financial stability, capability;
4. accurate determination of energy baseline (2- to 3-year average if possible) from which savings will be calculated; and/or
5. track record of potential contractor in matters, such as actual versus projected savings.

B. State the acceptable level of risk to the organization;

1. indicate planned procedures to ameliorate or manage the risks, and
2. determine the level of funds you are willing to pay out for someone else to assume certain risks. (Guaranteeing the savings, for example, may add 5 percent to the share the ESCO receives.)

IV. Indicate related energy services the organization needs; e.g., maintenance, energy accounting or preventive maintenance software, etc. Consider other capital needs or services where energy savings could "pay the freight."

V. Discuss the financial analysis procedures to be used for decision-making on energy projects; simple payback, life cycle costing, net present value, cash flow improvement, level of indebtedness and cost of delay.

VI. Discuss the best possible financing options:

A. Identify existing procedures; possible options;

B. Establish the range of acceptable rates and terms.

VII. Conclude with a statement as to the level of commitment management must be willing to make.

Setting Criteria

After working through the self-assessment guide, you will have established a framework for decision-making regarding your energy and financing needs. The remainder of this chapter and the two that follow are designed to help to secure the services to meet those needs through a performance contract. (Or, to provide ESCOs guidance in services and procedures that will enhance their proposals and presentations.)

Before any solicitation is issued, you should determine precisely what needs done, the kind of firm you want to do it, and how to determine whether or not a given firm is qualified to meet your stated needs.

Setting the criteria and weighting them will guide the solicitation process and form a framework for the ESCO's proposal preparation and your evaluation.

Some factors growing out of needs identified through the self-assessment process merit further consideration before solicitation criteria can be set.

Preliminary Considerations

Through the self-assessment and planning process, some internal decisions as to what is needed have been made. Other items to consider relate to the size of your facility and the extent of work for which energy financing is desired. These factors can help determine whether or not it is in your best interest to develop a larger package in cooperation with other firms or institutions in the area.

Collaboration

Some organizations are too small and/or too remote to be an economically viable customer for an energy service company. Or, the savings potential may not be enough for an ESCO to accept size and/or location draw-backs. In other cases, the services are offered, but the organization's share of the savings is relatively low due to these factors.

If this is a possibility in your case, you might wish to consider cooperating with other firms or institutions in the community to create a large enough package to attract an energy service company or to secure greater financial benefits. In such instances, the cooperating managers usually agree to use the same energy service company, but retain the right to negotiate their own contracts.

It is also possible for several small organizations that are part of a utility's customer base to be part of that utility's use of performance contracting for demand-side management (See Chapter 17.)

Buildings

For organizations with multiple buildings, a decision may need to be made regarding which buildings to include in a solicitation. Buildings scheduled to be closed within the probable life of a contract generally should not be included. Buildings scheduled for major renovations or additions can be included, but the base year will have to be negotiated separately.

Administrators are frequently inclined to try out performance contracting procedures on one or two of their largest energy consumers. This usually works against the organization's interests for several reasons; (1) a larger package will attract more service companies and more attention to innovative approaches, (2) it takes nearly as much administrative time to solicit and negotiate a contract on one or two buildings as the whole lot, (3) those buildings not included will continue to waste energy during the "pilot" process, and (4) it is a lot harder to get service on the more efficient buildings after the "cream of the crop" has been treated.

"Trying out" performance contracting will also suggest to the board, staff or public that performance contracting is in its experimental phase, which will tend to make them more cautious. The movement is no longer new or untested. The guidelines are in place and the industry has matured to the point where informed management can, and should, proceed with confidence.

If management/board insists on a pilot phase, a compromise may be reached wherein a certain level of success on the "treated" buildings provides the basis for a contingency to extend the contract to another group or all buildings.

Equipment vs. Full Service

The decision also has to be made as to whether or not the organization wishes to use vendor financing to buy specific new equipment, such as a computer-based energy management control system, or wants to look at all cost-effective measures that will save energy.

If the goal is to secure specific equipment with no front end cost, then vendor financing should be used. If the goal is to operate the total facility as effectively as possible; then, the service of an ESCO should be sought. This decision will influence the type of firms from which proposals will be requested, and the solicitation criteria that will be set.

Since a comprehensive energy package requires more services, the ESCO administrative time and detail is much greater. It usually requires an organization with an annual energy bill of at least $100,000 and strong energy savings potential to carry this administrative burden. Smaller organizations may, of necessity, opt for vendor financing, such as "paid from savings" equipment purchases.

Engineering Audits

As an added precaution, management sometimes may wish to obtain an independent engineer's audit prior to seeking a performance contract. This approach offers an informed basis for judging the proposers' technical qualifications. When this approach is used, it should be recognized that engineering services will be paid for twice, for ESCOs seldom, if ever, guarantee another engineer's work and will want to do their own audits.

Technical competence can sometimes be satisfactorily determined by checking the firm's track record. If an independent judgment seems advisable, you can ask for firms under final consideration to submit an audit representative of the caliber of work it will provide. This sample audit report can then be reviewed by your engineer.

In very large, unique or complex projects, you may want to ask for a test audit of one building, or a portion of a building, in a complex. If an organization is large enough to encourage energy service providers to risk such an investment, it may have to pay audit costs to the final contenders who do not receive the contract. Even this, however, is less expensive than the cost of having comprehensive audits performed on all the buildings and then paying for them again indirectly through the energy service company's "package" of services.

Once decisions have been made regarding these preliminary considerations, management is then ready to set the criteria for requesting services and to determine the weight each criterion should carry.

Establishing The Criteria

Energy financing criteria generally fall in some well established categories. Within these categories, the criteria can be modified or embellished. They should be set up to meet the organization's finance and energy needs and to enhance proposal evaluation procedures. Major energy financing concerns have been addressed in the categories presented below. The list, however, is not intended to be all inclusive and should be modified to address all concerns a particular firm, utility or institution may have identified.

The stream of criteria presented below are not in any priority order. Weighting the criteria is discussed later.

Proposal Presentation

Format adherence, responsiveness to solicitation, clarity and completeness of information, documentation.

Firm's Qualifications

General experience and background in performance contracting. Working with the similar organizations and more particularly with similar facilities;

Type of project, equipment proposed. Any unique qualifications;

The proportion of its business the firm has devoted to performance contracting and the percentage of work that has been done in similar facilities;

Financial resources/stability—adequacy of financing, firm's ability to back up guarantees over the length of the contract;

Resources available/offered other than financial—including computer-aided design, test/checkout, training, support services;

Reliance on and quality of joint venture partners, subcontractors;

Demonstrated ability—record, references, quality of references; and

Local support mechanisms—maintenance and emergency response provisions.

Key Personnel

Almost as critical as the qualifications of the firm is the background and experience of key personnel assigned to the project, particularly the project manager. Major emphasis should also be placed on the qualifications of the site manager, design engineer, and financial manager.

Key personnel qualifications may be placed in the firm's qualifications category, but a judgment regarding the credentials and experience of those *actually assigned to the project* needs to be made independently.

The percentage of time the key personnel will devote to your project is also a key factor. Big names with great expertise that are never seen after the contract is signed seldom help get the job done.

Management Schedule

Organizational lines of authority;

Ability to communicate, coordinate and manage projects;

Ability to mobilize the team, take corrective action as needed and complete project expeditiously;

Prompt project implementation;

Management record with like projects; and

Quality assurances.

Technical Approach

Quality and comprehensiveness of the sample audit (or test audit if requested); depth of recommended energy conservation opportunities; estimated savings;

Limitations—longest single payback or combined payback the firm will consider; other conditions;

Techniques and equipment used for testing and/or simulation;

Ability to identify specific actions to be taken; exploit new technological opportunities;

Technical expertise of the assigned personnel; particularly in similar facilities.

Equipment of preference—selection and flexibility; manufacturer's warranties; local representation; integration into complete system. (If ESCO is a manufacturer or locked into certain lines of equipment, appropriate application may be limited.)

Monitoring, field support capability;

Maintenance provisions on installed equipment; maintenance on existing equipment and additional costs for this maintenance;

Respect for the patient, learning, work environment—intrusion, comfort parameters; and

Range of services.

Financial Benefit to The Organization

Projected level of *total* energy savings: cash-flow potential to the organization; demand-side management for utilities; level of investment in *capital equipment* (NOT total dollars invested);

Extent of guarantees—your risk exposure is reduced if savings are guaranteed to exceed debt service;

Guarantees to cover any desired or required maintenance fees (You need to pay special attention here as the cost of required maintenance may not be under savings guarantee. It pays to ask potential ESCOs very directly what required costs are not covered with the guarantee.)

Buy out, contract continuation, and termination provisions;

Formula—baseline determination procedures, savings, billing calculations; demand costs; variables—weather, occupancy; explicitness and fairness of methodology; assumptions underlying formula;

Treatment of energy price fluctuations; risk to organization if a floor price is required;

Insurance - liability, etc.;

Performance and payment bonds needed or provided;

Audit fee for non-awarded firms if required;

Innovative financing opportunities; e.g., when energy cost savings don't accrue to organization, or are diverted to county or state treasury; what services and/or equipment (such as maintenance or classroom computers) are offered in lieu of cash;

Net benefit to organization; i.e., value of installed capital modification equipment and positive cash flow; percentage ESCO retains for services and guarantees.

Weighting The Criteria

Some factors are more important to a given organization than others. For example, *Proposal Presentation* and *Firm's Qualifications* are not of equal significance over the life of the contract. Giving a little weight to "Proposal Presentation" encourages careful compliance and quality presentations that expedite proposal evaluation. The firm's qualifications, on the other hand, are usually the determining factor in contract award as well as contract implementation and, therefore, would naturally be given more weight.

Some evaluators weight the projected cost savings very high; however, one should be leery of the smoke and mirrors approach. Such projections are not of much value unless the firm has demonstrated the ability to meet, or even exceed, projections made to other customers in the past. If projections run much higher than other proposers, it pays to do a background check.

Key Personnel might be weighted separately or as part of the Firm's Qualifications and/or under technical expertise as part of the Technical Approach.

CHAPTER 9
ENERGY FINANCING OPTIONS

Since energy efficiency is essentially an economic issue, the financial aspect of energy management is central to an effective program. The money side of energy has three major aspects: (1) buying energy and accounting for its use and cost; (2) determining relative cost-effectiveness of potential energy saving measures; and (3) finding the financial resources to do the job. This chapter focuses on the third aspect, financial resources and, more specifically, the options available within performance contracting.

The most important message in this book to owners, business managers, financial officers and facility managers is: *there is money available for economically sound energy conservation measures.* No one in business, industry, government or non-profit organizations should put off retrofits due to the lack of funds. You need only review the options and select the approach that best meets your needs.

Owners and managers should think of energy management as a fiscal resource. Purchasing agents wouldn't last long in an organization if every year they bought 25 percent more copy paper than was needed. As the paper stacked up, eyebrows and questions would be raised. Yet, most organizations buy roughly 25 percent more energy than needed year after year... after year.

It's worth repeating: **"Energy efficiency is an investment, not an expense."** It is an investment that can generate real dollars for an organization. The very fact that companies offering performance contracts exist indicates that money can be made from correcting other people's inefficiencies. If energy services companies can do this, then it's good business for an organization to do so.

When budgeting systems permit, money saved from energy efficient O&M efforts can often be pooled to provide the resources for small capital investments. Savings from small capital modifications will often yield enough money to go on to larger efforts. In this manner, a positive cash flow situation can gradually fund energy efficiency improvements, *provided* that other budgetary demands don't claim the dollars first. The roll over time also has cost of delay implications. While waiting to bank roll bigger efforts, the cost of wasted energy mounts.

Securing Financial Resources

Internal Clearance

Gaining internal clearance is the first step in securing energy funds. The technical staff may have ideas on how to cut operating costs, but somehow their thinking doesn't get far enough up the line.

For internal consideration and the required approval, energy project needs should be presented in a concise format—and in a language that decision-makers understand. Whether this presentation is going to the boss, to top management or the board, it should be kept simple, straightforward and devoid of energy jargon. If it is to be persuasive, some aspect must appeal to the decision-makers interests and needs. If the hospital can't attract the caliber of doctors it wants without upgrading the equipment, show how energy conservation measures might free dollars to lease an MRI.

The form prepared to "sell" your project should generally include several basic elements. The headings should include building identification, project title, a name of the person who prepared the summary and the date. The summary should address the nature of the work, costs, projected savings, cost-effectiveness and method of calculation, energy-related and/or non-energy benefits. To prompt closure, the form should have sign-off slots as appropriate. The form in Figure 9-1 on the following page can be adapted to meet a variety of organizational needs.

Local Support

Once the specific measures have been cleared internally, some organizations, such as public schools, may need to seek local voter authority to raise the millage, sell bonds, or borrow money. Even if a school or hospital receives a federal energy grant, as discussed in Chapter 20, some matching funds are necessary.

While capital improvements for the sake of saving "energy" may not win voter approval, sound business investments to save *money* will. Sharing energy program results in a routine fashion paves the way for further discussions and investments. Simple statements, which describe efforts that have already been made and the financial benefits resulting from these actions, have proved very successful in gaining local support. With this groundwork, future projects and their expected return on investment, can be discussed. Chapter 13 offers specific suggestions for effective communications strategies.

Federal and State Grants

Federal and state grants are available to various types of organizations for a variety of purposes. Some grants offer assistance with energy audits; others will provide support in purchasing specific types of equipment; i.e., solar collectors, lamps. To find out what grants are available, the stipulated purposes and conditions of the grants, eligibility and application procedures, a call to the state energy office should provide some answers. If a state does not have an energy office per se, staff personnel in the *local* state legislator's office or the congressman's office may be helpful in ferreting out appropriate grants.

Calculating the Cost of Delay in securing grant funds should be part of the decision process. If it will take a year or more to obtain the grant funds, the lost savings potential begins to add up and the "free" grants may be too costly. Money at 8 percent interest, for example, may become quite attractive if you face the loss of two years savings potential with an ROI of 30 percent.

ESCOs should also do their research as to the availability of these grants, as the funds from a grant may leverage the package so that more

PROPOSED ENERGY CONSERVATION OPPORTUNITIES

Building Name ______________________________

Address ______________________ Phone ______________

Project Title ______________________________

Date ____________ Prepared by ______________________

Approved by ____________________ Date ______________

PROJECT DESCRIPTION ______________________________

ENERGY INFORMATION: Current usage, demand, present cost, projected costs. ______________________________

COST AVOIDANCE/SAVINGS POTENTIAL ______________

METHOD OF CALCULATING COST-EFFECTIVENESS ________

CAPITAL INVESTMENT NEEDED ______________________

Offset value in life-cycle costing ____________________

NON-ENERGY CONSIDERATIONS ____________________

ACTION TAKEN ____________________ DATE ____________

Figure 9-1. Sample Project Proposal Form

ways to help a customer save energy are possible. The ESCO should also be prepared to help the potential customer compare the net financial benefit of a grant and its waiting period to the use of other financial mechanisms where funding is more immediate. The funds available and the typical payback level in a state should also be ascertained; for, even after a long wait, a grant may not be awarded.

Utility Rebates

As environmental pressures mount on utilities to manage their energy demand as an alternative to creating more generating capacity, technical and financial assistance to utility customers has become more common. In states where rate reform allows utilities an incentive to conserve as well as produce electricity, this support is even more extensive.

As discussed in Chapters 16 and 17, utilities themselves are turning to performance contractors to provide technical and financial support to their customers. Some utilities have become painfully aware that operations-dependent rebates; e.g., HVAC, do not necessarily reduce demand. ESCO's on the other hand guarantee the reduction. In performance contracting cases, a local organization becomes a "host" for reducing the utilities capacity needs. Arrangements vary, but the host generally gains some benefit from the utility's involvement in the process.

Both ESCOs and organizations seeking technical and financial support should keep abreast of local utility offerings and interests. Support may be confined to rebates or run the gamut from audits to energy efficiency design, to dollars for retrofits. The utility support may be all an organization wants at the time, or it may become a key piece in a performance contract, broadening the services an ESCO can offer.

A financial analysis of utility support should be made. If only lamp rebates, for example, are offered the end result may be some "cream skimming" that ultimately denies the organization the in-depth services of an ESCO. That is, the lighting retrofit may skim enough savings potential off the top to remove an organization as a viable ESCO customer. Or, the use of the rebates independent of the ESCO arrangement, may prevent the performance contractor from offering other services that would provide much needed equipment and modifications, saving the organization far more money in the long run than the value of the rebates.

The Energy Savings Business

The whole concept of performance contracting is based on a sound precept: **ENERGY CONSERVATION MAKES MONEY.** In a properly financed energy efficiency project, energy cost savings can exceed debt-service payments. This additional saving can generate immediate positive cash flow. In effect, an organization can "sell" future energy savings for immediate capital improvements, services and/or a positive cash flow.

Because cost control of operations can be developed as part of sound fiscal management, energy efficiency improvements are increasingly viewed as investments with attractive returns. As a result, outside investors are making retrofit capital available to organizations through various financing procedures, and are sharing the energy costs savings with local governments, retail outlets, schools, hospitals, businesses, colleges, universities, states or the federal government.

As discussed in Chapter 1, there are a number of terms used to describe private sector energy financing. Three terms are used interchangeably in many instances: alternative financing, savings-based arrangements and performance contracting. These three terms actually address different facets of private sector energy financing. *Alternative financing* suggests that there are other options than the traditional sources of revenue; *savings-based arrangements* indicate that in order to access the non-traditional sources, energy savings are shared in some fashion; and *performance contracting* refers to the responsibility of the energy service company to perform to specified standards. Since the standards and services to be performed are generally addressed in the contract, this facet is discussed in more detail in the chapter on contracts.

The term "performance contracting" is increasingly being used to cover all aspects of private sector energy financing, and is used in this book to include all savings-based arrangements and most non-traditional funding mechanisms. Occasionally, performance contracting is used to stress that the services offered are more than "shared savings." This is confusing, however, as performance contracting is a rubric for many financing mechanisms, including shared savings.

Performance contracting is one of the most promising areas of energy efficiency financing. Because performance contracting is still a relatively young industry, procedures are still evolving. While there are many different approaches to performance contracting, all are based on the premise that the customer and the private company will split the savings, or avoided utility costs, resulting from improved energy efficiency.

Standards are becoming more established. As the performance contracting industry has matured, procedures have actually become more flexible. If your organization's needs and terms make good business sense, ESCOs exist that will meet those needs. Unfortunately, as in any new field, not all firms have the credentials that assure quality work. On the plus side, the guarantee aspect tends to weed out the undesirables faster than in other industries.

Performance Contracting Opportunities

In the face of capital constraints, the management of any organization can take an aggressive approach to developing their energy cost savings potential by using performance contracting. Accessing private capital to improve energy efficiency can:

- generate immediate positive cash flow;
- relieve the tax burden of funding from state and local budgets for government entities;
- avoid the cost of delay which is incurred through lost savings potential and efforts to maintain old inefficient equipment while seeking traditional financing sources;
- make use of private sector energy management expertise;
- capitalize on the opportunity value of the organization's money not diverted to energy investments; and
- provide for needed capital improvements paid for out of energy savings.

Performance Contracting Cautions/Difficulties

Even if performance contracting seems desirable, you still need to weigh a number of factors before proceeding. To facilitate analysis, the concerns have been grouped into four categories: financial, procedural, technical, and legal.

Financial

The same financial benefits that accrue to the energy service company could accrue to the business, school, hospital, local government, or state agency in almost every instance IF—and it's a big IF—the organization has the initial capital, professional staff and management expertise to do the job effectively and promptly, and to stay on top of the project over the years. You must weigh the costs of doing the work in-house against their potential share of the avoided utility costs. Factors to consider include the:

- in-house expertise, including engineering and monitoring capabilities; manpower available
- maintenance benefits;
- opportunity value of the money;
- avoided indebtedness and voter/legislative authority required by some public entities to incur debt;
- risk management; and
- other services which energy service companies could offer.

Also included in this financial analysis should be:

- whether your organization is interested in equipment ownership, or building modifications that require ownership;
- the availability of capital;
- cash flow conditions, including cost of delay;
- economic projections and estimates of savings; and
- organizational risk; e.g., the degree to which you are required to make payments, take equipment ownership, assume maintenance responsibilities, etc.

Technical

Technical considerations which should be assessed include:

- existing condition of energy consuming systems, viability of modifications or replacement, and the capability of the equipment to be more tightly controlled;
- to what extent does the planned life of the facility warrant the work;
- the potential for cost-effective structural modifications; code implications;
- sophistication of current controls and potential benefit of a microprocessor-based energy management system installation or upgrade;
- quality of on-staff, or consulting energy engineering, services available to the organization—for design and troubleshooting; and
- the energy-related technical expertise of operations and maintenance staff.

Procedures

Procedural concerns include:

- local preferences and/or procurement requirements regarding bidding, Request for Qualifications, Request for Proposals procedures;
- means to evaluate bidders and their proposals;
- payment formulas;
- remedies by owner; e.g., repairs, improper maintenance;
- insurance, guarantees, maintenance; and
- termination and buy out provisions.

Legal

Sources of legal constraints that could impact the financing procurement include:

- building and safety codes;
- conflicts with other contracts; e.g., existing maintenance contracts;
- state legal constraints and procedural requirements; i.e., multi-year obligations, accounting procedures, etc.;
- indoor air quality implications;
- the degree to which control over a operations can be relinquished to a third party in the case of public buildings; and
- owner's liability.

It should be noted here that performance contracting delays are often incurred when attorneys are asked to review the contract. These delays can be significantly reduced if attorneys are asked to be involved early in the process.

Financing Sources and Options

Private capital is available to finance energy efficiency improvements in industrial, commercial and institutional buildings through a number of sources, which offer various financing arrangements. As you deliberate, the sources you might choose should be based upon the financing structure that most appropriately meets your objectives. The primary performance contracting sources are energy service companies, vendors and design professionals.

Energy service companies. The term, energy service company (ESCO) refers to any firm that provides the services and/or financing you might require to implement comprehensive energy cost reduction measures, usually in a performance-based arrangement. The firm generally provides the identification (audit), design, installation, and the financing of energy conservation measures (ECMs) for the customer. The ESCO is typically not a manufacturer of any specific equipment and is experienced in determining the appropriate application of a range of equipment of various types and makes. The ESCO usually ensures the performance of the installed ECMs by guaranteeing that the savings generated from measures are sufficient to cover all the related annual debt service. Although there are many different financing structures offered, they generally meet the criteria of a guaranteed savings, shared savings or a combination of the two. Further variations arise from the tax-exempt status of the customer and when the customer takes title of the equipment; i.e., lease, installment/purchase, direct purchase.

Guaranteed savings - This arrangement establishes a fixed payment that satisfies the debt service requirements for the installed equipment and building modification and the ESCOs fees. The ESCO guarantees that the energy savings will cover the customer's obligation. Sometimes the agreement includes maintenance on existing energy-related equipment or other services, which are outside the guarantee.

Ownership of the improvements may pass to the organization at the beginning of the lease. The investor typically retains a first security interest (usually title) in those improvements until the loan is paid.

While performance contracting is used as a rubric for all nontraditional financing mechanisms, you should be aware that performance contracting is also used in lieu of "guaranteed savings" to distinguish the guaranteed savings approach from shared savings.

Guaranteed savings through tax-exempt installment/purchase This method of financing typically provides financing for nonprofit institutions via a tax-exempt installment/purchase (frequently referred to as a municipal lease); however, distinctions between the two in terms of multi-year payment obligations are sometimes made. An ESCO will typically guarantee that the savings generated from the improvements will cover all of the annual installment payments to be made. Institutional administrators in most states can satisfy multi-year liability conditions by entering into one-year contracts with automatic contract renewal and standard non-appropriation language. The tax-exempt installment/ purchase is structured like a simple loan, except that the interest income to the lessor is tax-exempt.

This arrangement is among the most popular, as the obligation is not usually included in the institution's financial statements as a liability, and it provides an attractive, low, tax-exempt interest cost, only slightly higher than a bond rate.

Shared savings - In a shared savings arrangement, payments to the ESCO are predicated on a percentage of the energy savings generated from the equipment installed. The ESCO assesses the customer's energy savings potential and provides the capital for cost-effective measures. The organization and the ESCO share in the avoided costs on an established percentage basis negotiated at the time of the contract.

Because the ESCO and the customer usually share in the savings in some fashion, the popular press tends to use the term "shared savings" as a rubric for most performance contracting arrangements. Those involved in energy financing generally refer to *shared savings* as a particular financial arrangement in which equipment is leased or purchased through a percentage of the savings. In this book, and increasingly in the energy financing industry, the term "shared savings" is used in its narrowest interpretation.

The term "paid from savings," popular in the Southeast, may be a type of shared savings, or it may be a type of "vendor financing" used to obtain certain equipment, discussed later in this chapter.

At the end of the shared savings contract period, ESCOs may retain ownership of the equipment. If the customer does not take ownership of the equipment, this financial loss contributes to shared savings typically being the most costly of the performance contracting approaches. Occasionally an organization may prefer this procedure, particularly with energy management systems, as it can get the latest state-of-the-art equipment with contract renewal. If the ESCO continues to own the equipment at the end of the contract and the customer wishes to continue to use it, the ESCO will generally enter into a subsequent agreement offering a split in the avoided costs that is more favorable to the organization.

Since payment varies with the level of savings and the *price of energy*, the firm takes the greatest risk in this approach. Consequently, the cost of financing is much higher. Therefore, more investment funds go into financing and less into energy efficient equipment. All other conditions being equal, the accepted payback periods are apt to be cut by about one-third compared to guaranteed savings. As a result projects, which are judged cost-effective under other financing arrangements, may not be funded with shared savings financing at the same level of investment. This risk/financing/payback interaction forces an almost exclusive implementation of short-term payback items; this "cream skimming" does not work in the customer's best interests. Since the ESCO takes such a risk and it is seldom in the customers best interests, shared savings is less attractive and less apt to be used except as a way to finance specific equipment and, in the case of federal agencies where the law specifies "shared savings."

Some utilities, new to the business, have used "shared savings" because they view it as risk free. Others have not fully appreciated the distinction between shared savings and other performance contracting options.

It is not unusual to find a combination of the installment purchase and shared savings. In such situations, the savings in excess of that needed to guarantee the debt service may be shared between the customer and the ESCO (although it may go entirely to the customer). This is popularly referred to as a municipal lease with a shared savings wrap-around. As a wrap-around, shared savings does not carry the financing burden described above. Sharing the "extra" savings motivates both parties to stay on top of things.

Vendor Financing

By definition, a vendor in the energy financing industry, is a manufacturer of a specific brand of equipment. This type of financing, therefore, is primarily used to obtain specific energy efficient equipment. The vendor typically guarantees to the customer that the equipment payments will not be greater than the energy savings generated from it.

In some instances, vendor financing is viewed as a lease/purchase agreement rather than an incurred debt.

This approach offers only limited assistance in reducing energy costs, as it focuses only on the equipment being purchased and does not encompass a consideration of other energy conserving opportunities.

Vendors, on occasion, may act as an ESCO and offer a broader range of services. For example, Johnson Controls and Honeywell may sell their controls equipment on a "paid from savings basis." At other times, these companies provide comprehensive audits and supply a range of equipment and services through subcontractors. In such instances, they should be considered as an ESCO source. Since the audit is done by the equipment seller, even when the vendor acts as an ESCO, a bias in favor of certain equipment may exist. On the other hand, performance contracting can be viewed as a means a company uses to back up its product and services.

Vendor financing frequently uses a shared savings approach with the vendor receiving 80 to 90 percent of the savings. As noted above, paid-from-savings, is a variation of shared savings and may require that up to 100 percent of the savings goes to the vendor. This approach is most attractive to the customer when the brand of equipment has already been determined, but the capital is not available for direct purchase.

Design Professionals

Increasingly, architects and engineers who offer technical assistance to improve energy efficiency also arrange financing. These financing arrangements may seek the services of an ESCO as described above.

Some design professionals take their fee as a portion of savings generated. Others defer their fee or have the client reimbursed for their fee by the ESCO. Since design professionals may associate their fee with the avoided costs or with the financing arrangement, the term "energy service provider" is sometimes used to encompass ESCOs and design professionals.

There is an inherent potential for a conflict of interest if the engineer, as designer, also selects the ESCO or specific equipment. The customer should look to the engineer for specifications, but reserve the right to select the equipment.

You should always retain the right to select the ESCO. Some engineering firms offer ESCO solicitation consultation; however, unless the engineering firm has a long history of successful contracts using this approach, it is wise to secure independent consultation for this service. One engineering firm in Georgia, for example, didn't recognize the importance of a re-open clause much to the later dismay and financial loss of its customers.

As noted in Chapter 8, customers sometimes feel that a full engineering study of a facility will give the organization some key guidance. ESCOs will seldom guarantee performance based on the recommendations and calculations of an unknown engineer. The ESCO will do its own audit and it will be part of the project costs. The organization will, therefore, pay for two audits on the same facility. The technical security the organization wants can be achieved more economically by using the engineer as a technical adviser in the solicitation process as discussed in Chapter 8.

As administrators have become better informed about performance contracting opportunities, they have started to structure their own financing arrangements. As a result, the distinctions among the options discussed above have blurred and hybrids have emerged. The tax-exempt installment/purchase with a shared savings wraparound discussed earlier is one type of hybrid.

Pluses and Minuses in Energy Service and Financing Mechanisms

This section offers a way to compare the six basic service/financing mechanisms, which are used most frequently to finance energy projects.

- I General obligation bonds (for institutions)
- II Installment/purchase
- III Tax-exempt municipal lease/purchase
- IV Vendor (manufacturer) financing
- V ESCO shared savings
- VI ESCO guaranteed savings

The major advantages/disadvantages for the six financing mechanisms are listed on the following pages to assist you, or your energy-related professional, in determining the energy financing option that best meets your organization's needs. As presented, the mechanisms are generic in nature and not intended to reflect the specific services or financing offered by a particular firm. The "Pluses and Minuses" are not intended in any priority order and may, in fact, vary in importance in different situations.

You are cautioned that there may be local conditions or organizational needs/opportunities not listed that may, in fact, outweigh all other considerations. For example, a university may have an endowment fund earning interest at seven percent. Should calculations show that the endowment fund could earn 25 percent, if the university invested in its own energy saving potential; then, the administration would be wise to use its own funds rather than any of the financing mechanisms listed. The services provided by an ESCO might, however, still be desirable.

I. GENERAL OBLIGATION BONDS

(For Institutions)

PLUSES	MINUSES
1. Lowest possible interest rate	1. Time delays; bid & spec; attorney involvement
2. Long-term repayment schedule	2. Large upfront $$ required a. bid & spec - engineers b. attorneys c. administrative time
3. Institution owns equipment	3. Maximum risk - no equipment performance guarantees
4. Traditional "bid & spec" approach	4. No savings guarantee
5. Traditional financing	5. Institutions seldom predicate action on specific economic ties to a project; i.e.,payback, cash flow, account ability
	6. Debt on financial statement; affects on bond rating

7. Financially rigid
   a. no reserve funds
   b. only fixed interest rate
   c. level payments required
   d. no interim financing
   e. no future addition

8. No on-going monitoring

9. No added expertise or outside organizational momentum

II. TAX-EXEMPT INSTALLMENT/PURCHASE
(For Non-Profit Organizations)

PLUSES

1. Lowest interest rate compared to bonds

2. Long-term repayment schedule

3. Organization owns equipment; performance guarantees

4. Flexible
   a. reserve funds allowed
   b. rates
   c. interim financing
   d. level or graduated payments
   e. future additions, changes accommodated

MINUSES

1. Time delays; bid & spec; attorney involvement

2. Large upfront $$ required
   a. engineering specs
   b. attorneys
   c. administrative time

3. Required payment risk; no equipment performance

4. Without ESCO ties to savings, organization does not gain as much in support expertise and organizational momentum

5. Buyout option as a loan	5. No on-going monitoring
6. Traditional "bid/spec" approach	6. Not traditional; not well understood by most attorneys
	7. No savings guarantees
	8. Debt appears on balance sheet

III. TAX-EXEMPT MUNICIPAL LEASE/PURCHASE

All 6 of the above plus	First 7 of the above, in addition
7. No debt on balance sheet	8. The uncertainty posed by "non-appropriations" language increases interest rates

IV. VENDOR FINANCING
(Used to Purchase Specific Equipment)

PLUSES	MINUSES
1. No debt on balance sheet	1. Limited equipment approach
2. Performance guarantees provided	2. Higher interest rate (not tax-exempt)
3. Economic ties to project	3. Ownership of equipment not usually provided (may have buyout provision)
4. May provide monitoring, maintenance support for equipment purchased	4. Short term payment schedule
5. Low up-front engineering costs	5. Limited $ available

6. Relatively quick implementation	6. Typically includes required service contract
7. Limited risk	7. Not traditional financing
8. Provides interim financing	8. Limited future additions to financing package
9. Added expertise	9. Reliance on vendor
10. ESCO profit motive	10. Limited flexibility
11. Can negotiate	11. Not a comprehensive energy management approach
	12. Usually no debt purchase option

V. ESCO SHARED SAVINGS - SERVICE AGREEMENT

PLUSES	MINUSES
1. Limited risk (no save = no pay)	1. Limited $$ available (restricted by return on investment and risk concerns)
2. On-going monitoring	2. High interest rates; high risk; not tax-exempt
3. Comprehensive equipment approach	3. Equipment ownership not usually provided (may have buyout provision)
4. Low upfront engineering costs required	4. Short-term payback schedule
5. Relatively quick implementation	5. Future additions may be limited

6. No debt on balance sheet	6. Non-traditional financing
7. Some flexibility	7. Dependency on ESCO
8. Added expertise and organizational support	8. Low return to organization
9. ESCO profit motive	
10. Reliance on ESCO	
11. Provides interim financing	
12. Can negotiate	

VI. ESCO GUARANTEED SAVINGS WITH TAX-EXEMPT INSTALLMENT PURCHASE

PLUSES	MINUSES
All advantages of installment purchase except tax exemption, plus	1. Financial commitment on provided lease
7. Performance guarantees	2. Not traditional financing
8. Economic ties to project	3. Reliance on ESCO
9. On-going monitoring	4. Usually some operating controls held by ESCO; (organization must stipulate temperature/lighting conditions to assure comfortable productive environment)
10. Low upfront engineering costs required	

11. Relatively quick implementation
12. Limited risk
13. Comprehensive equipment approach
14. Greater flexibility
15. Maximum return to organization vis a vis other performance contracting mechanisms
16. Added expertise and organizational momentum
17. ESCO profit motive
18. Can negotiate equity build-up
19. Provides interim financing

5. Subjective method of selecting ESCO
6. ESCO requires role in specing/installing equipment
7. Baseline/saving calculations must be clearly established/understood
8. If the organization has the funds, expertise and accepts the risks, it could keep the money it would otherwise spend for financing, services, and guarantees

Figure 9-2. Financing Options Relative Benefits

ENERGY SERVICE/FINANCING MECHANISMS

Degree to which conditions apply signified by how completely the circle is filled in.	GO BONDS	INSTALL PURCH.	TAX-EX LEASE	VENDOR FINANC-ING	ESCO SHARE SVGS.	ESCO GUAR. SVGS.
ENERGY MGMT. BENEFITS						
Comprehensive energy mgmt. approach	○	○	○	○	●	●
Prompt implementation	○	○	○	●	●	●
Firm has economic tie to savings	○	○	○	●	●	●
ADMINISTRATIVE MATTERS						
Risk - savings guar.	○	○	○	●	●	●
Risk - equipment perf. guar.	◑	◑	◑	●	●	●
Title to equipment	●	●	●	◑	○	●
Traditional "bid/spec"	●	●	●	◑	○	◑
Organization holds operation control	●	●	●	◑	◑	◑
Equity build up	●	●	◑	◑	◑	●
Buyout options	○	●	●	◕	◔	●

No reliance on outside party	●	◕	◕	○	○	○
SERVICES						
Gain outside expertise	○	○	○	●	●	●
Monitoring - perf. & savings	○	○	○	●	●	●
Maintenance provided	○	○	○	◕	◕	●
Service contract not required	●	●	●	○	○	○
FINANCE CONSIDERATIONS						
Low interest rate	◔	○	◕	○	○	◑
Flexible finance structures	○	●	●	○	○	●
Long term financing	●	●	●	○	○	●
Few upfront dollars required	○	○	○	◕	●	●
Interim financing usually provided	○	○	○	●	●	●
No effect on indebtedness; bond rating	○	○	●	●	●	●
ESCO profit motive	○	○	○	●	●	●
Usually best perf. contracting				◑	◑	●

CHAPTER 10

EVALUATION PROCEDURES

The evaluation procedures set forth in this chapter are presented from the perspective of the organization seeking performance contracting services. While it offers guidance to management, ESCOs can glean some ideas as to how they might be evaluated and how they might make their proposals more effective.

In actual practice, the process of evaluating proposals will follow the solicitation procedures discussed in the next chapter. The intended evaluation procedure precedes a discussion of solicitations because evaluation criteria and process concerns should influence the way the request for qualifications and consequent responses are developed. As Abraham Lincoln once observed, "If we could first know where we are and whither we are tending, we could better judge what to do and how to do it."

Evaluating the qualifications of an energy service company (ESCO) usually requires a multiple disciplinary approach, including technical and financial expertise. The evaluation process, therefore, typically involves a committee. Unless the organization soliciting an ESCO's services has in-house performance contracting experience, it pays to support the committee's deliberations with a consultant. ESCOs, of course, are apt to tell you the services of a consultant are not necessary.

You may opt for performance contracting, because you do not have the funds to do the needed work on your own. A frugal posture may cause you to resist hiring a consultant. Unless someone in the organization is thoroughly familiar with performance contracting, this is an exceedingly expensive way to "save money." The costs of a performance contract consultant, and/or an engineer as a technical consultant, can be assigned to the project, and the costs covered by future savings.

Establishing Evaluation Procedures

Prior to structuring the evaluation process, the committee should meet and agree on definitions, scoring and procedures. Preferably, the criteria, weighting, etc. will be set by the committee before the solicitation is issued. The solicitation, such as a Request for Qualifications, will inform potential proposers of the services required and the criteria that will be used to judge their qualifications. (Assessing the organizations needs and setting criteria are discussed in Chapter 8.)

Establishing Criteria and Weighting

To establish consistency among evaluators, scoring procedures need to be determined at the outset.

Proposals can be scored on a 0 to 10 scale for each criteria and are entered on an evaluation form, usually a worksheet for each of the major criteria. Those scores are then transferred to a summary sheet. At this point, the total score for each major criterion is multiplied by the agreed upon weighting to reflect the relative importance of the criterion. The sum of the weighted scores for each criterion provides the total score.

The scores are usually based on a frame of reference as follows:

(0): Criterion was not addressed in the proposal or the material presented was totally without merit.
(1): Bare minimum.
(2): Criterion was addressed minimally, but indicated little capability or awareness of the area.
(3): Intermediate Score between 2 and 4.
(4): Criterion was addressed minimally, but indicated some capability.
(5): Intermediate Score between 4 and 6.
(6): Criterion was addressed adequately. Overall, a basic capability.
(7): Intermediate Score between 6 and 8.
(8): Criterion was addressed well. The response indicates some superior features.
(9): Intermediate Score between 8 and 10.
(10): Criterion was addressed in superior fashion, indicating excellent, or outstanding, capabilities.

Worksheets for each criterion can be broken out into factors to be considered. A detailed listing of these factors makes sure the evaluator considers each aspect—and that the proposer has not omitted, whether purposefully or inadvertently, any important information. Examples of the above criteria applied to team qualifications are shown below.

CRITERION: QUALIFICATIONS OF THE PROPOSING TEAM

FACTORS TO BE CONSIDERED:

a) Experience of the prime contractor with previous projects of similar size and type.

b) Experience of the joint venture partner (or suggested subcontractors in the team) with previous projects of similar size and type relative to their stated special expertise.

c) Experience of the proposed project manager as it relates to this project, as well as the qualifications of the assistant project manager; site manager; financial specialist; engineers for design; etc. Percentage of time suggested personnel will devote to the proposed project.

d) Resources (other than financial) available to the team for computer-aided design, equipment fabrication, test/checkout, on-site assembly/erection, training, etc.

SCORE:

10 = Fully-qualified and experienced personnel; comprehensive facilities experience, particularly qualified to perform designated duties, sufficient time allotted, excellent resources.

8 = Generally experienced personnel; allotted time adequate, some unique facilities, adequate resources.

6 = A majority of the personnel proposed are experienced in the general fields required; adequate resources; allotted time barely adequate.

4 = Personnel proposed appear competent in their fields; some resources are available; allotted time marginal.

2 = Some personnel proposed appear marginally capable; resources appear limited; allotted time questionable.

0 = Personnel appear inexperienced; resources appear insufficient for the job; allotted time inadequate.

CRITERION: ADEQUACY OF FINANCIAL ARRANGEMENT AND THE NET PRESENT VALUE OF COST SAVINGS TO THE ORGANIZATION

FACTORS TO BE CONSIDERED:

a) Soundness of the estimate of the cost of services and equipment for the work proposed.

b) The degree to which the cost savings are guaranteed and the form of the guarantee.

c) Degree to which cost savings are based on measurable quantities, projections of baseline values or estimated quantities.

d) Terms of the contract provide optimum benefit to the organization (length, return on investment, payment schedule, share of savings, etc.)

e) Ability to finance project.

SCORE:

10 = Firm (or investor partner, joint venture partner, or subcontractor) has indicated that extensive capital and cash flow resources are available; terms of the contract are extremely favorable; the firm proposes the highest net present value of cost savings to the organization among proposals received; the savings are fully guaranteed; and the organization risks no financial exposure.

8 = Financial resources are more than adequate: terms of the contract are favorable; the firm proposes a reasonable net present value of cost savings and are based on quantitative measurements, reasonably solid estimates, or projections of baseline values; the savings are partially guaranteed and organizational risks are minimal.

6 = Financial resources appear adequate: terms of the agreement are adequate; the proposal offers the second or third highest amount of cost savings; the savings may be guaranteed or are reasonably sure of being achieved; and the organizational risks are limited.

4 = Financial arrangement and resources appear marginal; net present value of cost savings is low; and the organizational risks are of some concern.

2 = Financial arrangement and resources appear inadequate.

0 = There appear to be no financial resources for the firm; no cost savings are given in the proposal; or they are not likely to be achieved.

CRITERION: TECHNICAL PERFORMANCE ESTIMATE

FACTORS TO BE CONSIDERED:

a) Adequacy of the equipment to provide the services proposed and its integration into a complete system.

b) Extent to which the proposed system will interface adequately with the electric/gas/steam utility, and with the existing facility energy system/equipment.

c) Adequacy of the proposed operation and maintenance concept, including training of facility personnel.

d) Degree to which the proposed system meets all environmental requirements wherever applicable.

SCORE:

10 = Equipment proposed is state-of-the-art and has proven reliability. The interfaces are well described and appear fully understood. O&M and training is well described and completely adequate. Environmental regulations will be met or exceeded in all regards (where applicable). Altogether, the system design is superior.

8 = Altogether, the system design appears to be better than average. The equipment appears to be reliable. The interfaces are competently described and adequate, as is the O&M training. Environmental regulations will be met.

6 = The overall system design appears to be average. The descriptions of the equipment performance and reliability, interfaces, O&M training, and environmental regulations are included but not fully documented.

4 = The overall system design appears to be marginal. The descriptions of the equipment, interfaces, O&M, training and environment are sketchy or omit important details.

2 = The overall system design appears to be inadequate to ensure continued performance. There are major omissions in the descriptions.

0 = The system proposed is completely inappropriate.

CRITERION: MANAGEMENT, SCHEDULE AND QUALITY ASSURANCE

FACTORS TO BE CONSIDERED:

a) The organizational structure is clear and well-defined; the lines of communication are direct; and the management appears to be well informed and responsive.

b) The proposed schedule appears to be reasonably rapid without being either too tight or dilatory, and allows for reasonable meshing of parallel and sequential activities.

c) The proposed management structure and quality assurance program can identify quality control problems promptly, and take effective remedial action.

d) The proposal clearly defines the organization's supplied resources; the proposed resources are reasonable for the project.

SCORE:

10 = The proposal indicates a very effective management structure, has satisfactory scheduling, and offers the best quality control among the proposals received.

8 = The proposal indicates an above average degree of effective management, prudent scheduling and quality control.

6 = The proposal indicates a good degree of effective management, prudent scheduling and quality control.

4 = The proposed management, schedule and/or quality control appear to be below average.

2 = The proposed management, schedule and/or quality control appear to be marginal.

0 = The proposed management, schedule and/or quality control appear to be inadequate.

The committee, with the support of a consultant, can enlarge upon or modify these guidelines and develop other criteria so that factors specific to your organization's needs are evaluated.

Evaluation Instruments

Two different formats are offered for consideration. The first format utilizes an outline with a straight forward mathematical approach for weighting the criterion. A summary sheet and a "Typical Financial Benefits" worksheet for this approach are shown in Figures 10-1 and 10-2. The summary sheet and supporting worksheet presented here can be adapted to guide those wishing to use this approach.

Since the details in a proposal seldom fit preconceived molds, the decision-matrices in the second format provides greater flexibility. The approach shown in Figures 10-3, 10-4, and 10-5 offer more flexibility to highlight particularly attractive features or some strongly held reservations. This decision-matrix summary sheet and the supporting worksheets allows the evaluator to view and compare at a glance the ways each firm treated certain factors. For more complex projects, the decision-matrix approach is preferred.

The criterion and weightings suggested in these figures are just that; suggestions. They tend to reflect actual practice, but every organization needs to decide the relative importance of certain criterion. For example, the weight of "5" for the proposal presentation is established to encourage proposers to follow the format prescribed in the RFP to facilitate evaluation. The object of the effort, however, is to select a firm, not a proposal. If too much weight is placed on the proposal presentation, the process could defeat the purpose.

CHECKING REFERENCES

As part of the evaluation process you should always ask for, *and check*, a potential ESCO's references regarding work they have done in similar facilities. The references cited by an ESCO, of course, are usually the best they have to offer. Therefore, it pays to dig deeper on the finalists.

Several sources are available to the enterprising procurement office. Colleagues are always good sources. The state energy office in the ESCO's home state may be a viable source. The state personnel cannot endorse a private firm and are not apt to rule out any firm; however, listening to what they "don't say" can help. If they suggest you contact people who have used the proposer's services, do it. Indirectly, you'll find out in a hurry whether the energy office is high on a certain firm, or otherwise.

PROPOSAL EVALUATION FORM—SUMMARY SHEET
(Mathematical Approach)

Evaluation Criteria*	Weight*	Score		Points
Proposal Presentation	5	× ______	=	________
Firm's Qualifications	35	× ______	=	________
Key Personnel Qualifications	10	× ______	=	________
Technical Approach	20	× ______	=	________
Management, Schedule, Quality Assurances	10	× ______	=	________
Financial Benefits	20	× ______	=	________
	100%			— — —
Total Points				________

Comments: ______________________________

____

____

____

____

______________ Proposer

______________ Evaluator

______________ Date

*Criteria and weights are given as examples only.

Figure 10-1. Proposal Evaluation—Mathematical Approach—Summary

FINANCIAL BENEFITS

Energy Baseline and Savings Methodology

1. What are the data sources and inputs that will be used in computing baseline and savings? How will it be collected?

2. How are the following factors to be accounted for in the firm's computations:

   consumption

   rates/prices

   weather

   occupancy and hours

   equipment changes

   end-use conditions

   other

3. Clarity of methodology and sample calculation for computing:
   a) baseline
   b) savings

4. Percentage of savings retained by ESCO ________%

Figure 10-2. Proposal Evaluation—Mathematical Approach Worksheet

Decision-Matrix — Summary Sheet

Criteria/Firm				
Proposal Presentation	__ x 5 = __	__ x 5 = __	__ x 5 = __	__ x 5 = __
Firm's Qualifications	__ x 35 = __	__ x 35 = __	__ x 35 = __	__ x 35 = __
Technical/ Service	__ x 20 = __	__ x 20 = __	__ x 20 = __	__ x 20 = __
Management	__ x 10 = __	__ x 10 = __	__ x 10 = __	__ x 10 = __
Financial Benefit	__ x 30 = __	__ x 30 = __	__ x 30 = __	__ x 30 = __
Comments: Evaluator:	Total	Total	Total	Total

Figure 10-3. Proposal Evaluation Decision Matrix Summary Sheet

Decision – Matrix = Financial Benefit

CRITERIA/FIRM				
PROJECTED LEVEL OF TOTAL ENERGY SAVINGS				
ORGANIZATION'S SHARE (% of savings)				
INNOVATIVE ENERGY FINANCING —Payment Schedules —Interim Construction Financing				
CONTRACT YEARS & RELATION TO —Savings —Services —Other Benefits —Combined Payback				
FORMULA Establishing Baseline Billing Calculations Demand charges				
TREATMENT OF VARIABLES —Occupancy —Weather —Energy prices —Price floor?				

(%) LEVEL OF INVESTMENT IN CAPITAL EQUIPMENT & MODIFICATIONS				
EXPLICITNESS AND FAIRNESS OF METHODOLOGIES				
RISK EXPOSURE DISTRICT REQUIREMENTS —Performance, payment bonds —Insurance —Operational control —Guarantees				
PROJECT TERMINATION —Buyout provisions —Return to original status				
PROJECT CONTINUATION —Subsequent contract benefits —Renewal —Maintenance				
COMMENTS EVALUATOR ______________	TOTAL	TOTAL	TOTAL	TOTAL

Figure 9-4. Proposal Evaluation Financial Benefit Worksheet

If you still remain uncertain as to the potential performance of a proposer, members of the committee can resort to the ultimate check-up: ask the competition in the area to identify any project where other proposing firms did not perform as promised.

CONTRACTOR SELECTION

While not essential, it is usually valuable to have the committee meet again after independent judgments have been made and submitted. This removes the possibility that committee members misunderstood the instructions, allowed biases to intrude, or overlooked a key area that might be brought out by another committee member during the discussion.

After the contractor has been selected by the committee, it is generally necessary to have top management or a board ratify this action.

This ratification is generally followed by a letter of intent to the selected firm. This letter notifies the contractor of his/her firm's selection and stipulates the time frame and conditions for contract negotiations.

Typically, a master contract is then established. Specific recommendations, associated maintenance costs, etc. for specific buildings or complexes are treated in addenda or schedules. These procedures are more fully discussed in Chapter 12.

CHAPTER 11

MINDING YOUR P'S AND Q'S

Proposals that cover all facets of energy financing and services are cumbersome to evaluate and expensive for an energy financing firm to prepare. The time and cost of preparing a major proposal is apt to discourage some energy service providers; i.e., energy service companies (ESCOs) or registered professionals, from submitting a proposal. This is especially true if the savings opportunities are limited, or the administration or delivery of services relatively burdensome. A cumbersome solicitation process, therefore, limits the options available to smaller organizations.

The rules for soliciting proposals are pretty simple:

Keep it short!

Keep it open!

Keep it simple! and

Get only the information you truly need to make an evaluation!

Keep it short! It takes time for your people to prepare big requests for proposals (RFPs). And you don't do yourself any favors by doing so. The only excuse for extensive RFPs is in the case of federal and some state government efforts where acquisition regulations compound the problem. Most RFPs (outside of government) should not exceed 30 pages, plus any appendices.

Keep it open! Detailed specifications are a throw back to "bid/spec." Performance contracting solicitations should be based on eliciting information as to a firm's qualifications; not price. (There are some exceptions, such as conditions under which utilities are unfortunately presently soliciting proposals.) Describe what results you want; not how to do it. Detailed specifications eliminate options ESCOs can offer. Your organization may lose out on all the rich experiences ESCOs can make available to you. It's their *business* to identify all cost-effective energy conserving opportunities; not yours.

Precise specifications not only put a box around the ESCO's opportunity to help; it may spell out a project in such a way that the ESCO cannot guarantee the savings.

Keep it Simple! Some RFPs raise the question, "Who are you trying to impress?" Or, "You want what!?!" Your organization is more apt to get clean, direct proposals if you issue clean, forthright RFPs.

The obligations imposed on responding ESCOs should be kept within the realm of reason. An organization with 100+ buildings recently issued a *preliminary* request for qualifications that required the responding ESCOs to audit some facilities and to guarantee a level of savings for *all* its buildings. At this preliminary stage, the idea is absurd and serves no purpose! Since it is a large organization, some ESCOs did offer qualified responses in the hope that the organization would get more realistic as the process progressed.

So you can better understand how ESCOs react to RFPs and RFQs that go beyond acceptable conditions, an internal memo from a national ESCO regarding this particular RFP is presented in the figure on the next page. The identification of the ESCO, the issuing organization, and some numbers have been changed to preserve anonymity. The essence of the memo remains intact.

Getting The Best

The more you ask for; the more tedious and costly the evaluation process. Most "kitchen sink" RFPs are a testimonial to organizations, who have not truly thought through their needs and/or the performance contracting process. Or, managers who must satisfy so many internal masters that the primary purpose of the document is all but forgotten. In "many masters" situations, the use of oral interviews for the pre-qualified few should be considered.

The "get what you need" admonition cuts both ways. ESCOs

XYZ ESCO
SOMEWHERE, USA

MEMO

TO: TOP MANAGEMENT

FROM: REGIONAL REP

DATE: ____________, 1991

SUBJECT: ABC Junior College
RFP FOR ENERGY MANAGEMENT SERVICES

On ______, 1991, I received notice of the ABC Jr. College's intent to solicit proposals for energy management services (copy attached). I attended the bidders conference on _____, 1991. The list of attendees is also attached for your information.

Several bits of information discovered during the meeting have lead me to the decision to not bid on the project. My reasons are as follows:

> The request for quotation requires that bidders conduct detailed energy audits on 450,000 to 520,000 (4 or 5 facilities) sq. ft. of ABC's facilities, to be submitted with the bid. An effort of this type would cost approximately $25,000 in consulting engineering or XYZ's contractors fees, OR, consume 50% to 75% of my time for two months. My budget and staff limitations don't allow consideration of either option.

The request for quotation requires that bidders guaranty savings at the proposal stage for 10.3 million sq. ft. of facilities, based on the results of the audits on the sample 4-5 facilities provided with the bid. It's not in XYZ's best interest to take the risk on making a promise of savings based on the results of a detailed look at a 4.5% sample of the building inventory.

The building inventory to be addressed is already operating very efficiently (69,200 BTU/GSF/YR and $.82/GSF/YR in 1990). Therefore, meeting the five year performance contracting requirement will be very difficult.

FIGURE 11-1. An ESCO's View of a Cumbersome RFP

should be constrained by format and length. Unless it's one of those government things, one must worry about an ESCO that can't tell you who they are, what they can do and how they package their program in 50 pages or less.

To Test or Not to Test

ESCOs, who repeatedly succumb to test audit conditions, have to recover their costs somewhere. The test audit costs run up the overhead. Ultimately, those costs must be borne by the customers—particularly those organizations that asked for test audits.

Unique organizations with unusual operations or peculiar problems may warrant a test audit. Others should resist this cumbersome embellishment if they possibly can. (The buildings at "ABC Junior College" in Figure 11-1 are standard; no exceptions. A test audit was totally unwarranted.)

The major difference between a full blown RFP and what has become known as a request for qualifications (RFQ) is typically the test audit requirement.

As discussed in the previous chapter, historically the best way to judge the technical capability and approach of an energy service company is to have an energy engineer review an audit report of the organization's test site. Use of test audits was more important when the industry was young and the abilities of individual firms were untested. Today, established ESCOs have proven track records and an increasing number of organizations are accepting well documented references/ case studies in lieu of on-site test audits. A test audit is just one more obstacle that may prevent a well-qualified ESCO, who could meet your organization's needs, from even proposing. Audits are costly and ESCOs are increasingly reluctant to audit on speculation.

At the very least, audits should only be requested of a short list of pre-qualified ESCOs. This saves the ESCOs time, improves their odds and reduces the burden on your organization that would result from the evaluation of a large number of test audits.

An attractive alternative is to ask proposers to submit an audit for a similar facility, which is representative of the work they propose to perform. This sample audit can still be evaluated by your engineer and can also be used as a basis for a representative presentation of an ESCO's financial approach.

Requests for Qualifications

Since the firm's qualifications are the most important single criterion, there is a growing inclination to engage in some type of preliminary merit selection or brief request for qualifications (RFQ) procedure prior to any in depth consideration of technical competence. If your organizations is inclined to use this approach, exercise care that this preliminary screening does not yield qualified firms that look good on paper, but in practice offer inexperienced teams to do the work.

The RFQ procedure narrows the focus to a select group of candidates. A second stage, request for technical proposals, oral interviews, and/or sample audits may then follow. This phase of the selection process should reach beyond a firm's general qualifications to the individuals who will be assigned to the project.

Several years ago, Mr. William H. Ferguson presented a performance contracting case study used by a Rhode Island state agency, which effectively summarized the advantages of the RFQ process. Ferguson described the relative merits, stating:

The objective of the purchasing process are to:

1. Maximize competition
2. Maximize choice of approaches
3. Minimize risk to the state

1. Maximize Competition. Prepare a list of several qualified firms from which to solicit proposals. These should be firms which have been subjected to preliminary investigation to determine that they are suitable for the project. Do not overtax the purchasing process by seeking too many proposals. Having too many proposals is also somewhat unfair to vendors as they put much work into proposal preparation and their chance of winning becomes slimmer. They may not want to bid on future projects if the competition is too keen. Ideally a minimum of three and a maximum of six is a good range. Using a two phase approach is best. The first phase would screen firms based on qualifications (the Request for Qualifications of the RFP process). The second phase would involve soliciting proposals from these prequalified firms.

2. Maximize Choice of Approaches. Do not over specify the project. If you attempt to specify too much, you may unintentionally rule

out some viable approaches to the project. Define the results rather than the means. Results are things such as: reduced energy consumption, positive cash flow, guaranteed minimum level of savings, maintenance of all installed equipment, specific comfort levels, lighting levels etc. Let the bidder define the approach. This is where he is an expert not the customer.

3. Minimize Risks to the State. The most important aspect to assuring your risks are minimized is to be certain the firm is financially stable and professionally qualified. This must take into account the financial stability of the firm, its experience, and the input of references provided by the company. In the case of the State's contract with CHPC [Citizens Heat and Power Corporation] we also required minimum guaranteed energy savings, minimum comfort standards, energy savings insurance, performance bonding of all subcontractors to CHPC, and liability insurance covering bodily injury and property damage. Many other provisions of the contract prevent the State's and CHPC's exposure to risks such as arbitration procedures, conditions and remedies of default, and base year adjustment procedures.

   The bidding process itself helps to minimize risks by providing a basis of comparison to assure that proposals are realistic.

The emphasis Ferguson places on the RFQ process as a means of maximizing competition should help allay management concerns for those who are used to the sealed bid approach and somewhat apprehensive as to whether or not the RFQ process will provide the desired level of competitive bidding.

Ferguson's second point should become the first law of RFQ writing: *the more you ask for the less you get*. Defining the results rather than the means will reduce the burden of RFQ preparation and encourage more innovative responses.

Merit Selection

For many years, organizations have secured architectural and engineering (A&E) services for design work under the Brooks Act, which provides for A&E firms to submit their qualifications on standard forms 254 and 255. The Federal Acquisition Regulations have been revised to incorporate professional services other than the improvement to real property. Several years ago *Building Design & Construction* discussed the implications for the revised rule.

A/B WORK DEFINITION CLARIFIED

> The Federal Acquisition Regulation has been revised to more fully define architect/engineer services in accordance with the Brooks Act.
>
> The new definition follows that of the previous regulation but now also incorporates professional services that do not specifically involve improvement of real property.
>
> The new definition helps to clarify who can perform services that don't involve the design of physical facilities, such as in the case of an energy study or feasibility study. Earlier regulations covered such situations, but regulations adopted more than a year ago did not.
>
> "Now we have a definition that's an improvement over the old definition in the regulation," said Milton Lunch, legal advisor for the National Society of Professional Engineers. "It states that the Brooks Act applies not only to design or improvements to real property, but also to other services that the contracting officer determines require the expertise of an architect or engineer."
>
> Lunch said that confusion has arisen when federal agencies have issued contracts that didn't involve the design of physical facilities. One of these "gray area" contracts was for analysis of the structural integrity of a building, but did not involve design services.

The expansion of acceptable procedures under the Brooks Act has prompted a discussion of using the 254/255 approach for selecting energy service and financing firms. It does offer a simpler acquisition alternative, which is already known to management and professionals offering technical support. For this approach to work, however, the forms would need to be modified so that qualifications pertaining expressly to energy service company requirements are incorporated. While expanding the A&E selection process to address ESCOs has been considered for a time, the use of a merit selection procedure has been sporadic and usually confined to pre-qualification procedures.

If a small organization has decided that it is desirable to seek cooperative arrangement with other entities in the community, prequalification of the proposers provides an agreed upon short list from which all the interested parties can work.

When performance contracting with the federal energy Institutional Conservation Program (ICP) grants is involved, special requirements must be met. If energy service companies are expected to deal with the additional complexities of matching a federal ICP grant, some preliminary selection process is warranted. The intricacies of using performance contracting as a match for the federal ICP grant are discussed in the last chapter.

Pre-Proposal Conference

The RFQ/RFP process frequently also involves a conference for prospective bidders, particularly for larger or complex projects. Much like a bidders' conference, the pre-proposal meeting is designed to clarify and expand upon the organization's needs. If it has not been made part of the issued RFQ, you should be prepared to describe the facility(ies) operation (hours, occupancy, processes, square footage, etc.) and offer utility records for at least one year, preferably two or three years. As part of the conference, proposers should be given an opportunity to walk through a building while staff respond to questions. Even though technical considerations may have been deferred, prospective proposers should have some opportunity to judge the organization's saving potential and whether the job potential meets *their* criteria.

Unless the project is very large, attendance requirements should be kept to a minimum, for this is just one more requirement that could discourage ESCO participation. Pre-proposal requirements put a burden on the industry and reduce the options for the end user; so required conferences should be well structured and substantive. If the project is awarded to an ESCO that did not attend a pre-proposal conference, then a real question as to the value of the pre-proposal conference must be raised.

A required pre-proposal conference, therefore, should be called only if there are some unique concerns or procedures that may substantially affect the proposals; e.g., the required procedures for using performance contracting as a match for federal energy grants as discussed in Chapter 20.

Unless the pre-proposal conference is a required part of the selection procedure, a firm's attendance is not, technically, a factor in making the selection. The conference does provide a subtle opportunity to size up the proposers, and for ESCOs to make a favorable impression. It is also an opportunity for an ESCO to size up a candidate for its services! Since performance contracting has been compared to a marriage, the pre-proposal conference might be viewed as a part of the "courting ritual."

The merit selection phase, whether by oral interviews, a standard qualifications form, or response to an RFQ, is the first step in the selection process. Figure 11-2 displays a flow chart of the major actions and decisions points in the selection process. Addenda or schedules are often added to the contract to treat the engineering studies of the facilities, particularly if the organization is large enough to warrant audits/installations done in phases. The flow of tasks may need to be modified if special conditions, such as an ICP grant, are involved.

As indicated in Figure 11-2, after selecting qualified firms, the organization can elect to hold oral interviews or issue an RFP. If oral interviews are held, the letter of invitation should outline the nature of the presentation expected and any other requirements to be placed on the proposer, such as a test audit. After the interview, you may wish to request a summary statement confirming or clarifying certain points raised during the meeting.

Request for Qualifications

The request for qualifications (RFQ) can stand alone in soliciting ESCO proposals, or it can be used to pre-qualify ESCO in a two phase selection process. In either case, the prospective contractors are asked to submit statements regarding their qualifications to undertake the project as described. The information requested should include the firm's background, designated employee credentials and experience; and a description of work for like clients, along with references. Energy cost and consumption data should be included in the RFQ so that responding firms can determine if energy cost savings potential is sufficiently attractive.

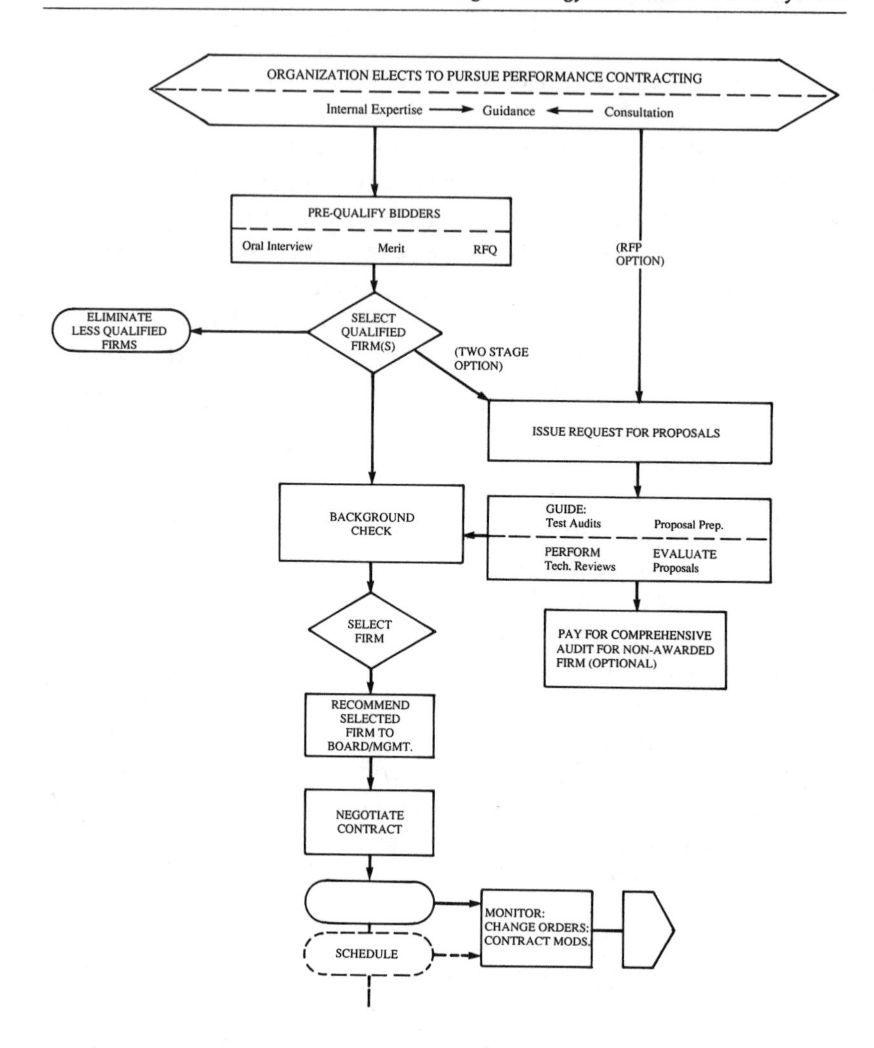

Figure 11-2. Major Steps in Energy Service Company Selection.

When used as the sole solicitation, the RFQ also seeks information on the ESCO's technical approach.

The organization then evaluates the qualification statements from the proposers. In some instances, no further information is required from the bidder for the end-users to make a decision. In the two phase procedure, this part of the selection process serves as a screening mechanism to be followed by limited competitive evaluation of their technical competence and the related financial benefit to the organization.

There are three disadvantages to relying on a RFQ/RFP two phase effort: (1) the qualifications proposal usually does not describe the firm's technical approach to the project, or the financial benefits to the organization, so you will not know for sure whether you have entered into the most favorable agreement (the most attractive opportunity, therefore, may be lost); (2) ESCOs may view the two phase submission as too burdensome, particularly from a smaller organization, and may be discouraged from proposing; and (3) when used with a technical proposal phase, it protracts the selection process, thus lengthening the time the organization must pay for wasted energy. A two-step selection process also lengthens the sales cycle, which ultimately adds to product/service costs.

In Figure 11-3, the basic elements of an RFQ are presented. A sample RFQ is presented in Appendix A. In using the language in the sample RFQ, you need to remember it only offers a starting point for developing an RFQ that meets your organization's particular needs. You are encouraged to modify it by expanding or deleting certain sections to reflect your precise needs.

Request for Proposals

Should an organization wish to use a request for proposal (RFP), the RFQ presented in Appendix A can be converted to an RFP by adding components that solicit information on the technical approach and associated financial calculations. Language eliciting technical competence should set the auditing and financial parameters. Should you decide to have a test audit done on given facilities, language similar to that placed at the end of Appendix A should be incorporated in the solicitation document.

Other Modifications

In converting the RFQ to an RFP, the document should be carefully reviewed for any cosmetic changes needed to make the solicitation consistent throughout. Other procedural changes, such as the time specified in the deadlines, should consider the longer time period required for ESCO to conduct a test audit.

Whatever process is used, the key to effective solicitation procedures is to ask for the information that will enable you to judge the proposers' qualifications and competence to meet the your organization's particular needs as established in the criteria.

REQUEST FOR QUALIFICATIONS

1. Purpose and scope—As briefly as possible, offer the over-all purpose of the project, the range of services you are looking for, and any limitations of importance to an ESCO.
2. Proposal procedures—submission information and expectations regarding the RFQ or later submission requirements; e.g., whether an audit will be expected from the qualified firms.
3. Pre-proposal—(bidders) conference information
4. Selection criteria and weighting
5. Contract requirements pertinent to costs to be incurred by contractor
6. Proposal format and content
   a. Contractor background and qualifications
   b. Trade references
   c. Personnel designated to participate in the project, their qualifications, and percentage of time each person will devote to the project
   d. Sub-trades performed and reliance on joint venture or subcontractors; qualification information on joint venture
   e. Prior relevant experience and references
   f. Annual report or audited financial statements for the ESCO's most recent fiscal year
   g. Demonstrated capability to finance the project
   h. Demonstrated level of performance bonding; professional liability insurance
7. Assurances, qualifications, limitations
8. Proposal format instructions

Figure 11-3. Request for Qualification Elements.

CHAPTER 12

CONTRACTS

Successful energy financing firms have learned that the most profitable energy service agreement, for both the firm and the end user, is one that rests on a sense of partnership. Successful ESCOs are willing to work with management and staff to arrive at mutually satisfactory arrangements.

Energy service agreements **are negotiable.** Owners and managers need not, in fact should not, "lose control" as discussed in the chapter on risk management. You can, *and should,* first determine the key elements the organization must have in an energy financing contract and the latitude within which you are willing to negotiate.

Understandably, an energy service provider has to have some assurance that it can protect its investment and can reasonably guarantee the savings. Contrary to fears engendered by some, this can be achieved without any negative impact on the internal environment. In fact the contract conditions can, and should, enhance the work environment.

Any firm that comes in with a "take-it-or-leave-it" contract and attitude is not a firm that will work *with* the you to achieve the best results.

As suggested in Chapter 11, a request for qualifications (RFQ) or a request for proposals (RFP), should ask for a copy of the ESCO's contract recently executed with a similar organization. As part of the final evaluation process, the contracts submitted with the proposals should be reviewed to get a sense of what the firm really expects. Contracts from the ESCOs in final considerations should also be reviewed by your organization's legal counsel. If your attorney does not normally provide counsel in contract law, it is prudent to seek additional or outside counsel.

Attorneys, not comfortable with the performance contracting concept, can be a major impediment to achieving an agreement. Since this type of contract may be without precedent in the attorney's experience, it will expedite the process if he or she is provided references of attorneys of record on successfully implemented contracts. The state energy office, the organization's consultant, or the ESCO can usually supply such information. It will also facilitate the process if the attorney is brought into the discussions early in the process.

Laying The Groundwork

Establishing the criteria, preparing the RFQ or RFP and evaluating the proposals lays much of the groundwork for the contract. Neither the solicitation document nor the proposal should be considered all inclusive *or binding*. It may be stated in the solicitation that you reserve the right to make the proposal part of the contract; however, all organizational conditions upon which the proposal was based should remain constant or the ESCO should not be expected to comply with this provision.

Items not in the solicitation or proposal can be placed on the table for discussion during negotiations. Modifications in what the organization asks for or the firm proposes to do are commonplace. Should the proposal or parts of it, by reference, become part of the contract, a statement should be included in the contract indicating that, in case of conflicting provisions, the contract prevails.

Contract Components

Topics generally addressed in energy service agreements are:

- financial terms and conditions;
- equipment/building modifications and services;
- user and ESCO responsibilities; and
- the construction contract provisions.

These items may all be covered in one contract or separate schedules pertaining to work in specific buildings, or clusters of buildings, and may be negotiated.

A non-profit organization may enter directly into a installment/ purchase agreement with the financier; so it can use its tax exempt status to obtain the equipment at a lower interest rate. In such cases, a parallel agreement with the ESCO can then be entered into to audit the facilities, install and maintain the equipment, provide other services and guarantee that savings will cover required payments.

As indicted earlier, performance contracting may also be referred to as guaranteed savings or alternative financing. The growing popularity of the term, performance contracting, rests on the stipulation in most contracts that the energy service company must *perform* to a certain standard (level of savings) as a condition of payment. These performance considerations are integral to the contract components and are implied throughout most contract provisions.

TYPICAL ENERGY SERVICE/FINANCING CONTRACT COMPONENTS

1. Recitals (traditional, but not essential)

2. Equipment considerations
   — ownership
   — useful life
   — installation
   — access
   — service and maintenance
   — standards of service
   — malfunctions and emergencies
   — upgrading or altering equipment
   — actions by end-user
   — damage to or destruction of equipment

3. Other rights related to ownership

4. Commencement date and term renewal provisions

5. Compensation and billing procedures baseline calculations baseline adjustment provisions, including a re-open clause

6. Late payment provisions

7. Energy usage records and data

8. Purchase options

9. Insurance

10. Taxes, licensing costs

11. Provisions for early termination
    — by organization
    — by firm
    — events and remedies
    — non-appropriations language (for government entities)

12. Conditions beyond the control of the firm

13. Default
    — by organization
    — by ESCO
    — events and remedies

14. Indemnification

15. Arbitration

16. Representations and warranties

17. Compliance with laws and standards of practice

18. Assignment

19. Additional terms
    — applicable law
    — complete agreement
    — no waiver
    — severability
    — further documents, schedules

20. Schedules (by designated group of buildings, or project phases)
    — description of premises; inventory of equipment
    — energy conservation measures to be performed
    — calculation of baseline
    — savings calculations; formulas

— projected compensations and guarantees
— comfort standards
— contractor O&M responsibilities
— O&M responsibilities of organization
— termination, default value, buyout option
— existing service agreements
— calculation of other savings; e.g., calculation of existing service/ maintenance contracts
— contractor training provisions
— construction schedule—approved vendors/equipment

A contract offered by a firm is designed to ensure that the ESCO's interest are protected. As in all contract negotiations, it is up to the customer to make sure its interests are protected as well. Prior to negotiating a contract, you will find it exceedingly helpful to consider the implications of the various key components and the latitudes within which an item can be negotiated. In other words, decide what is not negotiable, what conditions can serve as "trading stock" and in what priority.

Special contract considerations in the order in which the contract components are listed in the preceding pages are discussed below.

Key Contract Considerations

Equipment ownership. The financing scheme used and the point at which the organization takes ownership can affect the organization's net financial benefit. The useful life of the proposed equipment is a key factor in post-contract benefits.

ESCOs and/or their financiers usually insist on a first security interest in the installed equipment or collateral of equivalent value. You should make calculations and comparisons and determine how much latitude your organization has in this regard.

In the case of buy-out provisions, termination and default values, the capitalized equipment cost should be established in the original contract; terms, such as fair market value, need to be carefully defined.

Malfunction and emergencies. If the firm selected is not a local one, provisions for immediate, and back-up, service need to be spelled out. Local distributors for the selected equipment frequently serve this function with further back-up provided by the ESCO.

Maximum downtime needs to be considered. The allowable emergency response time will vary with the equipment installed and how essential it is to your operation.

ESCOs need to establish an understanding with the distributor or designated emergency service provider as to the timing and extent of emergency service before they are committed in the performance contract.

Firm actions, damage. Contracts proffered by ESCOs will discuss actions your organization might take that could have a negative effect on savings. The management needs to determine if these conditions are reasonable.

You also need to decide to what extent your organization should have the same protections.

The organization should also examine the redress available if actions taken by the firm have a negative impact.

Equipment selection and installation. You should reserve approval rights on selected equipment provided approval is not "unreasonably" withheld.

ESCOs must retain some rights if they are to guarantee the savings. Under some bid procedures, the energy service company may take on the role of a general contractor; writing specs, monitoring bid procedures, and overseeing installation. Again these may be services your organization needs, but they also serve to protect the ESCO's position on guarantees.

Contractual conditions used in any construction project; i.e., liabilities, OSHA compliance, clean up, performance bonds, etc., should apply.

Provisions for early termination. From the your organization's point of view, contract language regarding termination should include removal provisions, including length of time required and a provision for restoration of the facility to its condition at the time of contract origination or better.

ESCOs incur major exposure early in the contract for they incur the major expenses at this time and must depend on eventual savings to cover these costs. Buyout provisions must provide for ESCO recovery of costs incurred and a proportionate profit.

For further protection, ESCOs frequently specify that the organization cannot replace the equipment with equivalent equipment within a specified time frame.

Conditions beyond the control of the firm. Usual contract language absolves the ESCO of certain contract responsibilities under specified conditions. These conditions should be examined, and the merits of similar provisions for the organization should also be weighed.

Default language. Language frequently limits the conditions of default for the ESCO, but may leave it wide open for your organization. Since the financial burden is carried by the ESCO, this is not necessarily inappropriate. Similar language for the organization should be considered, however, especially if a nonprofit organization holds the debt service contract on the equipment.

Indemnification. Both the ESCO and your organization should be indemnified.

Assignment. You should insist on prior approval for any assignment changes of service responsibility or key personnel. Prior approval of subcontractors may also be desirable.

Applicable law. The ESCO typically presents a printed contract as the basis for agreement. The ESCO is apt to specify the applicable laws of the state in which it is incorporated. Should court action be necessary, this places an additional burden on your organization if you are located in a different state. Since applicable law provisions may just as easily specify your state, this provision may become "trading stock" in the negotiations process.

Calculation of baseline. Provision for calculating baseline should consider particularly mild or severe weather in recent years. Recent changes in the structure, building function, occupancy, etc. should also be factored into establishing a baseline. Recent O&M work to reduce consumption or any recent renovation which affects energy consumption should also be considered. Reopen language should provide for some adjustment beyond any agreed upon variations in climate, occupancy, etc.

Savings calculations formulas. This procedure is frequently made far more complex than it needs to be. One contract had nearly thirty pages devoted to the treatment of degree days! Weather or occupancy changes, added computers, etc., can affect savings; however, extensive contract language trying to anticipate every contingency only benefits the legal profession. The simplest way is to have a broad-based baseline and then agree to re-open, or negotiate, changes of greater than ± "X" percent.

The reduction in units of fuel and electricity multiplied by the current cost of energy by unit is the standard procedure for calculating cost of saved energy. Attribution of demand charge savings also need to be negotiated and included.

Changes in energy prices need to be considered: How is the burden of falling prices or the benefit of rising prices to be shared? If one party insists on a price floor, the other party should enjoy the benefits associated with increased prices.

The share of the savings will vary with the length of payback, the services delivered, the financing scheme selected, the risks assigned to the ESCO, the length of contract, and the like. The interrelationship of these factors needs to be considered in negotiating the your organization's share of the savings.

Comfort standards. The greatest fear employees associate with performance contracting is the loss of control of the work environment. The organization can, and should, establish contractually acceptable comfort conditions; e.g., temperature parameters, lighting levels, air exchange, degree of building level control needed (and override required) to assure a quality environment. The degree or latitude of control held by the organization can reduce savings. In such instances, the ESCO's risk becomes greater; so your organization's share will be less. The supposition that an energy financing firm will control the building operation is simply not warranted—unless management abrogates its responsibility and gives the control away.

Projected compensation and guarantees. The most attractive part of performance contracting is the idea that there is a third party out there who will make sure you can have new capital equipment *that works*, and you can enjoy a positive cash flow without any initial capital cost to the organization.

The manner in which the energy savings are guaranteed to cover debt service payments is a key component of a contract and deserves careful consideration. Since the quality of maintenance on energy consuming equipment affects savings, most ESCOs require related maintenance contracts. They may not, however, guarantee that energy savings will cover the required maintenance fee. If you regularly contract for maintenance and the ESCO's fee is not greater than the existing fee, this may not pose a problem.

A major reason for a contract is to identify and assign risks and provide appropriate recompense. The "guarantees" are the bottom line in making sure a contract works in your organization's favor. However, the greater the guarantees, or the risks shed, the lower the savings benefits will be. As discussed in Chapter 7, *money follows risk.*

As with any contract, your attorney should review the contract before signing. Through all the negotiations, frustrations and delays, it's well to remember that a good contract is essential to a successful project.

CHAPTER 13

GETTING RESULTS THROUGH EFFECTIVE COMMUNICATIONS

The ideas and concepts considered in previous chapters may be quite new to many of the people that you must turn to for energy efficiency and performance contracting support. Members of the governing board must understand what the administration is trying to accomplish, how it intends to go about it and what the results are apt to be. Operations and maintenance (O&M) staff must understand the "why" as well as the "what" of their role in energy efficiency, or the effort will fail. Care must also be exercised to assure O&M personnel that the new energy effort is not a criticism of their past efforts, and will, in fact, enhance their work. Building occupants need to know how they fit into the efficiency plan—and how they, in turn, will benefit. Without that understanding and the commitment to help, occupants can defeat the most sophisticated control system or elaborate management plan.

Once management starts the process of seeking an energy service company (ESCO) partner, communication becomes an essential component of the selection process. Once the contract is signed, regular progress meetings and close coordination become essential ingredients of an effective project.

There is nothing magical, nor particularly sophisticated, about the application of communications techniques to the various requirements of an energy efficiency program, or the performance contracting aspect of it. Effective managers know very well how to communicate with their boards, staff, clients, patients, customers or community. The absolute dependence of an energy efficiency program on the human element, however, makes a review of communications fundamentals valuable.

An energy efficiency project spends money on hardware, but whether energy and money is actually saved depends on people... and getting results depends on effective communications with the people involved.

But "energy," the buzz word of the 1970s, no longer works. In 1990-91, the Gulf War and escalating energy prices put "energy" back on the front page. Within a few months, concerns had once again faded and for most people "energy" today remains a not very interesting subject. Years of misinformation and cries of "crisis" have prompted apathy, resistance, and even skepticism.

The two real issues today that bolster the need to do energy efficiency work are **MONEY** and **ENVIRONMENT**.

We need to talk **MONEY** ... dollars saved rather than energy conserved. Everyone understands and is interested in money. Few would agree to throwing money out with the garbage; so you just might interest them in all you are burning up.

Cultivating and making use of that interest in relation to energy efficiency is the prime communication's challenge. In fact, if those dollars can be translated into a competitive advantage, computers, band uniforms, etc.—something of great interest to members of the organization "energy savings" become even more attractive.

A few years ago, Mr. Amil Akuja, energy program manager for the Los Angeles United School District described an effective communication's strategy used to gain support for a $15 million energy program. He said the new energy management program appealed to different decisionmakers for different reasons:

> We sold [the concept of installing an energy management system] to the schools' principals by saying the systems would improve the classroom environment. The board bought it because the systems would have a 3.5-year payback, and

the maintenance people bought it because with the alarm functions of the systems, their operations costs are lowered.

Understanding the art of communicating to the needs of specific groups, Akuja concluded that the program received support from all board members because it had saved the district $67 million in avoided energy costs over the previous ten years.

ENVIRONMENT—there is growing concern about the pollutants emitted by burning fossil fuels. The question of the 1970s was "do we have enough fossil fuels to sustain the economy?" The question for the 1990s and beyond is "Can we afford to burn what we have?"

Concerns about the outside air we breathe have now been compounded with worries about global climate change and the ozone layer. The Clean Air Act Amendments passed by the 101st Congress is still the biggest *energy* bill ever passed by Congress. The pressure this act exerts on the utilities to pursue demand-side management rather than create new generating capacity will help foster energy efficiency. The utilities are increasingly positioned to help customers cut energy consumption *and* to help tell the story of why reducing our energy use is a critical environment consideration. Performance contractors are actively supporting utilities and their customers in this endeavor. (Chapter 2, Color Us Green, has more material on the environmental implications of fossil fuel consumption.)

Effective energy management means that an administrator must understand and fully accept the critical role of communications in all phases of the effort. Effective communications is planned and orchestrated; it becomes as much a part of a successful program as does the work accomplished with a wrench or a screwdriver.

Communicating Energy Needs

The essence of effective communications lies in knowing: (1) the target audience... Who do you need to reach? (2) the purpose of the message... What action do you want to prompt? (3) What (and how much) do they need to know to achieve the results you want; and (4) What is the best time, route and format to use to reach your audience. How well these factors are understood and how they relate to each other will determine the degree of success (or failure) of any communication's effort.

Target audiences can, and must, be defined. At a minimum, they include top management, staff, building occupants. People in the community are a particularly important group to public institutions. If an organization is to function smoothly and effectively, none of these audiences can be neglected for long.

Bringing The Board on Board

No energy efficiency project can be truly effective without solid commitment; therefore, it makes sense to start this discussion of communications strategies with the board, the governing body or top management. Not all organizations have "boards"; however, for simplicity sake, the term *board* will be used in a generic sense to encompass all governing bodies and the higher echelon throughout this discussion.

The board needs to be informed regarding past energy management successes, and the projects that can be financed. If any part of that understanding is incomplete, getting backing for any new energy projects can be very difficult.

A few facts can help tell the story. Does the board know what energy costs per bed? Per pupil? Per square foot? Per unit of production? Do they have available information about the potential to save? Are utility matters a real part of the budget process, or just another line item that is passed over without discussion because "nothing can be done about it?"

A simple presentation to the board showing the money the organization is losing unnecessarily and the dollars that might be saved through improved energy efficiency is a good starting point. Try out some hypothetical (but reasonable) numbers, say 15 percent or 20 percent savings this year, and then project those savings for 5 or 10 years.

Then, ask for support to explore ways to make those savings possible.

Does the board know the environmental damage being caused by the burning of these unnecessary fossil fuels? The CO_2, NO_X and SO_2 emissions do not have to be calculated by each fuel exactly — some figures averaged to Btu reduction will get the message across.

A walk-through energy audit of several buildings will provide a good estimate of savings opportunities and suggest the steps needed to get the job done. This should also supply the basis for determining the dollar and environmental costs of doing nothing about the problem.

"Energy" information needs to be presented to the board in a straight forward manner. Burying board [or bored] members under discussions of Btu or elaborate discussions of variable air volume will only confuse the issue. A general discussion of what can be done, the dollars that can be saved and what those dollars could mean to the operation should be closely supported by environmental concerns.

Every CEO, president, superintendent, financial officer or chief administrator knows how quickly board members learn that anything anyone wants to do costs money. At this stage, it is valuable to introduce the idea that there are a variety of energy financing options available beyond traditional methods—and that several of these options do not require any capital investment or "new" money. Support to explore those options and develop a plan for their consideration comes next. Keeping up the flow of information and moving ahead steadily, but in rather well defined steps, will enhance the chance of success when it becomes time to "ask for the order" to use utility rebates and/or private sector financing.

On almost any board, there will be one or more members who want lots of details about almost everything. In energy, as in other areas, it pays to deliver those details, but only on request. Delivering supplementary materials or having additional discussion outside of board meetings may recruit some solid support, while not turning off those who really don't want to know that much about air intake volume, burner efficiency, or CO_2 emissions.

Performance contracting, differs rather widely from traditional financing methods. As every administrator knows, anything out of the ordinary, particularly when it comes to financing, may make board members uneasy. The ability to explain the options that are available with considerable clarity and confidence takes on added importance. An outside consultant may provide valuable guidance and communications support.

Citing others, who have successfully used performance contracting, may allay concerns. The fact that the federal government, state governments, hospitals, universities, school systems, etc. are all using performance contracting with increasing enthusiasm speaks to its effectiveness. The federal government, for example, has directed each branch of the military to enter into three performance contracts, called "shared energy savings" *each year*.

External Publics

Talking to external publics can be very important, especially for public entities. After all, the community is the ultimate boss, paying the taxes, and voting council members. It pays dividends to let the community know that the administration is doing something about energy costs and the environment. Audits and studies aimed at increased efficiency are news. Training programs to upgrade energy efficiency skills of operations and maintenance people are news when couched in terms of controlling energy costs and pollution while improving the learning, patient care, or work environment. Stories can be prepared for the local press, or additional opportunities to get the word out can be found.

When the time comes to move forward with a performance contracting project, the information about controlling energy costs and pollution without using local or state tax dollars is good business and good news for local governments, schools, etc. Proceeding in a business-like manner to do something positive about costs, improving cash flow and freeing up funds for special purposes without adding to the tax burden, is very welcome news indeed.

When a RFQ or RFP is issued, public organizations should let their external audience know about it. It reinforces the fact that something is being done to use resources efficiently. When contracts are signed, publicize the fact that a program is off and running. If it is appropriate, emphasize the portion of the work that will be accomplished by local subcontractors. Work for local companies is always good news in the community. If the project is being financed by an energy service company, it helps to coordinate the development of any news releases with them to assure consistency and accuracy.

ESCOs can be very helpful in supplying public information strategies that have worked with similar customers. A performance contractor can become the conduit to networking with other organizations with performance contracts. The beginning of work and the installation of new equipment can provide a "news hook.".. an excuse... for communications to all concerned that things are happening on the energy/dollar front. It provides still another change to build support for, and an understanding of, the energy efficiency program.

Consider photo opportunities. If the energy savings "bought" new band uniforms, show the students trying them on. If the savings paid for new computers for the college, show them being unloaded. If the savings paid for replanting the atrium at the nursing home, show the planting with residents helping or looking on. If the savings increased the donation your firm can give to the Salvation Army, get a picture of presenting the check—or better yet show the good things that were made possible with the money.

Bringing The Staff on Board

Almost certainly, successful communications with staff is the key to the mechanical and financial success of any energy efficiency project. If the staff, at all levels, understands how the energy effort will benefit them, they will generally support the program. If not, determined "resisters" can find ways to defeat any plan. It is important to use every opportunity to put across the message that the energy efficiency effort is not just to save energy (who cares) but to save dollars that could be going for other, much more desirable things.

That is why one portion of the needs assessment discussed in Chapter 8 is devoted to developing a "wish list" and calculating ways to show how energy savings can pave the way to getting computer systems, books or new equipment... or cutting product costs.

Building occupants also need to know that energy conservation doesn't mean bundling up and freezing during the winter, nor does it mean living with far too much heat in the summer. A well engineered and executed energy management program can enhance overall comfort levels. Of course, some buildings will never be totally comfortable no matter what is done to the HVAC system; so it pays to be cautious with any promises.

And don't forget that the environmental message is just as important to the staff—maybe more so.

The total building population is an "audience" that will experience the results of the program. If heating and cooling systems work better (or worse), what is perceived to be happening will make all the difference in their acceptance of changes that occur. If this audience knows and understands what is going on and feels that the results will benefit each of them respectively, fewer instances will be created where, for example, an instructor, aided by students, destroyed a thermostat with a high heeled shoe. Communications to this vital inside audience can be tailored to fit various segments of building populations, so everyone can be let in on the "secrets" of energy efficiency.

As an important side benefit, it has been demonstrated, over and over that effective, internal communications is one of the most effective external communications tools. Nurses talk to patients, staff to friends, students to parents and to other students, and the word spreads.

Operations and Maintenance. A critical first step with operations and maintenance (O&M) staff is to affirm the good things they have been doing. They must not view preparations for an energy efficiency program or a performance contract as a criticism of their work. Nor should it be viewed as a job threat. Performance contracting can free the staff to do other work that has been put off far too long.

O&M personnel should be active participants in the effort from the beginning and should receive a steady flow of information as the project moves forward. Even when the process becomes a matter of issuing requests for qualifications (RFQs) and contract writing, the O&M staff should be kept informed of progress. They have a direct, personal interest, because when the project actually "happens," it will happen to them. Their attitude will have a major impact on results—good or bad.

Getting The Word Out

Messages need to be kept simple. Most people are not interested in all the details of the project, but they do want to know how it affects them. So, tell building occupants what it all means in terms of comfort and money saved for other things. Tell O&M staff what the changes can mean in improved equipment to work with, perhaps fewer complaints from occupants, better levels of maintenance and control. Tell employees about comfort and dollars that can be salvaged from energy bills. Tell the board (and the public) that there are ways to increase energy efficiency, gain capital equipment, reduce environmental damage and increase cash flow without the commitment of organizational funds. Those who want all of the details will ask and should be accommodated, but for the most part the real question will be, "What does it mean for me?"

Communications are a key part of energy efficiency planning and should be considered a part of every phase. When the audits are done, publicize the fact, giving credit to those employees who do the work. When consumption drops in particular facilities, recognize the custodians at a board meeting.

Projects do not always go smoothly nor precisely follow the plan. Careful attention to communications throughout every phase provides the needed base of credibility if things should go wrong. And if they do, acknowledge problems while they are small and establish the fact that steps are being taken to deal with them. That way, there can be no sudden revelations of CRISIS. The keys to successful passage through the communications mine fields when problems arise are openness, candor, and adherence to the first (and best) rule of press relations: NEVER SPECULATE. If problems arise, never cut off the flow of information.

Getting The Job Done

Who should do all of the communications work? And, it is work. If the organization has an information director or public relations person, this effort should be a real part of his or her responsibility; but, it cannot be done alone. There must be a steady flow of information from those directly involved in the project to the person responsible for dissemination. This type of information effort often fails because someone is "too busy" to pass along details of what is happening and/or doesn't realize its communications importance. The information program will be effective only if it is considered as a regular part of the project and planned as carefully as engineering and finance. If communications is handled as an after thought, or "when I get around to it," there is a real risk of project failure.

The day when everything is finally in place and running well is a good time to celebrate. It is time to get the word out yet again. A time to review what has been done and to look ahead to what all the changes can accomplish.. a time to "point with pride." It is also an opportunity to salute the staff and recognize those who have helped make it happen. Recognition ceremonies may be appropriate. More than one certificate has found its way to the boiler room wall. Also express appreciation to those who may have been inconvenienced during the project. It's a great time to say thanks.

And when you have results, SHOW THEM. Keep the picture opportunities in mind. If management can demonstrate that the energy efficiency program has saved money for those special items from the "wish list," tell the story. Installing new lab equipment, patient care equipment or new playground equipment in the city park that were paid for through energy savings has real visual impact.

Communications is the application of common sense to the distribution of information. Successful managers have learned to use the tools and techniques of communications in all sorts of circumstances. An energy efficiency program depends upon the application of the best of those skills if it is to get results. With a performance contract, the partnership is strengthened through planned communication. It puts dollars in the bank instead of pollutants in the atmosphere.

PART II

MAKING PERFORMANCE CONTRACTS WORK

Global descriptions of the performance contracting process, as presented in Part I, fail to reflect the differences in specific applications and the nuances unique to certain market sectors. To close the gap, Part II draws from the "how to" foundation presented in Part I and applies the procedures to three very unique markets; i.e., federal government, utilities and institutions. For each of these markets, the chapters in Part II present performance contracting from the end user's and the performance contractor's points of view.

Increasing energy efficiency in our federal government's operation is a particularly challenging opportunity. As one would expect, it is not without its problems and frustrations. Federal agencies that would like to pursue private sector energy efficiency financing and the contractor offering the service must comply with laws, energy savings contract regulations and Federal Acquisition Regulations. (Although Congressional legislation may soon ease this burden). To illustrate the opportunities and the problems, Section 1 has two chapters. The first chapter provides insight from within the federal government by Mr. J. Terry Radigan, who at the time of writing, was the Acting Director of the U.S. Department of Energy's Federal Energy Management Program office (FEMP). FEMP is charged with providing energy expertise, support and guidance to the federal departments and independent agencies, so the federal government can reduce energy consumption. An important part of FEMP's service is aiding these departments and agencies in securing private sector financing and services to meet their conservation goals.

From the time Mr. Radigan joined FEMP in 1985, he carried the primary responsibility for guiding federal agencies in their use of shared energy savings.

Mr. Thomas K. Dreessen, President of CES/Way, the firm which holds the most federal energy savings contracts to date, shares some special industry insights from the contractor's perspective related to serving the federal government's energy and financing needs. At the time of this writing, Mr. Dreessen also served as President of the National Association of Energy Service Companies (NAESCO). Mr. Dreessen's remarks, however, are his personal views and not necessarily those of NAESCO.

Demand-side management (DSM), least cost utility planning and provisions in the Clean Air Act have placed increasing emphasis on energy efficiency as a resource utilities can use instead of incurring the costs of providing more generating capacity. In addition to time-of-use schedules and rebates to meet these needs, utilities are turning to energy service companies to augment their conservation efforts. Since the utilities consume 36 percent of the primary source energy in the United States, this market is becoming a critical area for performance contractors. At the same time, it allows utilities to use the resources of an ESCO to meet their DSM goals.

In Chapter 16, the opportunities and problems related to the utilities reliance on performance contractors for DSM are presented by Mr. David R. Wolcott, with the New York State Energy Research and Development Administration. From its earliest days, Mr. Wolcott has been active in performance contracting and has been instrumental in the research and development of its use to help state and local governments and institutions meet their energy needs. A pioneer in the use of performance contracting to meet demand-side management goals, he has offered guidance to utilities and ESCOs.

The ESCO perspective in the use of performance contracting to meet utility DSM needs is provided in Chapter 17 by Mr. Cary Bullock. Mr. Bullock is president of KENETECH Energy Management, Inc. and he also serves as vice-president of NAESCO. Mr. Bullock views are his own and not necessarily those of NAESCO.

The growing financial concerns and energy needs of public institutions make performance contracting particularly attractive to schools, hospitals, local governments, colleges and universities. The rigidity and timing of their budget processes make them particularly vulnerable to energy price fluctuations. For the energy service industry, institutions represent an exciting opportunity to serve the market and indirectly the general population, but not without some frustration. In Section 3, the unique characteristics of this sector are explored using the schools, as the case in point. The concerns and options facing institutions considering performance contracting are presented as a case study of Charles County Public Schools in Chapter 18 by Joe Lavorgna, Assistant Superintendent for Supporting Services and James Smoyer, electrical engineer. Mr. Lavorgna is a professional engineer who guided the district through the total performance contract process. The viewpoint of the energy service company dedicated to serving this market is offered by Mr. R. Scott Holland, National Sales Manager, Schools and Colleges, for Honeywell. The final chapter discusses the benefits and problems related to using performance contracting as a match for the Institutional Conservation Program energy grants, which are available from the U.S. Department of Energy through state energy offices.

SECTION 1

THE FEDERAL GOVERNMENT

The federal government represents the biggest opportunity for performance contracting and, to date, the biggest disappointment.

The Office of Technology Assessment (OTA) for the U.S. Congress estimates that the government could cut its consumption by 25 percent through more energy efficient operations. Since the government's utility bill is nearly $4 billion, this puts about $1 billion dollars per year on the table for energy service companies (ESCOs) and the government to save. A lucrative opportunity exists for the government to generate nearly a billion per year in new "revenues," and for ESCOs to do themselves, government and the taxpayers a lot of good.

The Congress passed a law in 1985 that enabled federal agencies to enter into "shared energy savings" (SES) contracts. When asked in 1988 for an assessment of the opportunity this presented, OTA predicted there would be 30 SES contracts by the end of fiscal year 1988. Several years have passed and the government has less than 10 SES contracts.

The problems associated with shared savings (discussed in Chapter 9) plus federal procurement regulations have caused problems, but concerns related to federal government performance contracting are broader than that. Chapters 14 and 15 discuss the successes and the problems that have emerged. Mr. J. Terry Radigan, Acting Director of the Federal Energy Management Program, U.S. Department of Energy, provides some keen insights into the program as it has operated so far. As president of the company holding the most federal contracts to date,

Mr. Thomas K. Dreessen describes the exciting investment potential, addresses some concerns from an ESCO's perspective and offers some recommendations in Chapter 15. For federal agencies and those who seek to serve them, these two chapters are worthy of careful deliberation.

CHAPTER 14

FEDERAL ENERGY EFFICIENCY FINANCING FROM A GOVERNMENT PERSPECTIVE

J. Terry Radigan

THE FEDERAL OPPORTUNITY

Federal buildings offer fertile territory for energy service companies (ESCOs) with the resources, fortitude, patience and flexibility to meet government needs and procurement procedures. At the start of the 1990s, the Federal government was consuming approximately 1.9 quads of energy annually costing roughly $10 billion each year. Of this amount, $4 billion was spent in Federal facilities and another $4 billion was spent in energy subsidies for low-income households. In addition, $1.3 billion was paid by the U.S. Department of Housing and Urban Development (HUD) for public and Indian housing development.

The Federal opportunity is large and diverse. Federal facilities include 500,000 buildings that are owned or leased, including 51,000 commercial buildings, adding up to nearly two billion square feet. There are also 422,000 facilities for military family housing. The U.S. Department of Defense (DOD) owns about two-thirds of the total Federal domestic floor space. The U.S. Post Office ranks second with about 9 percent of the square footage. The next largest "owner" is the General

Services Administration, which owns, leases or manages many of the buildings housing Federal agencies.

Most of the Federal infrastructure is old and frequently inefficient. The opportunity to cut operating costs through increased energy efficiency is significant. Recent assessments of Federal facilities indicate that 25 percent of the energy currently being used in Federal buildings could be saved through cost-effective modifications to equipment and the building envelope.

Toward More Energy Efficient Operations

Since the 1970s, both Congress and the executive branch have strived to promote energy efficiency within Federal agencies. Successive pieces of legislation passed by Congress and Executive Orders signed by the President have incorporated the benefits of past experiences, new technologies and approaches.

In its report, *Energy Efficiency in the Federal Government,* the Office of Technology Assessment (OTA) summarized the main acts of Congress regarding Federal energy management legislation, and the key provisions. OTA's summary is presented in Table 14-1.

Of the legislative actions reported in Table 14-1, several are particularly germane to private sector energy financing. The Comprehensive Omnibus Budget Reconciliation Act (COBRA) of 1985 (P.L. 99-272) encouraged Federal agencies to seek private financing and to implement projects through "shared energy savings" contracts. The Federal Energy Management Improvement Act (FEMIA) of 1988 allowed agencies to retain a portion of the savings to do additional energy efficiency work, which provided some incentive to agencies.

The National Defense Authorization Acts (NDAA) of Fiscal Years 1989-1991 strengthened the incentive concept. NDAA of 1989 allowed commanders of military bases to use half of the first year savings for "welfare, morale or recreation activities" at the base. NDAA of 1990 extended this provision to the first five years. NDAA of 1991 expanded the provision to allow one-third of SES contract savings for additional energy efficiency measures and one-third for improving family housing at the base *or* for welfare and recreational activities. The 1991 NDAA also

Table 14-1. Federal Energy Management Legislation

Law	Purpose	Provisions for Federal Energy Management
EPCA 1975	Increase domestic energy supplies and availability; restrain energy demand; and prepare for energy emergencies.	Directs President to develop mandatory standards for agency energy efficiency procurement policies and develop and implement 10-year plan for energy conservation in Federal buildings. Requires that Federal vehicle fleet meet corporate average fleet efficiency standards.
DOEOA 1977	Establishes department of energy to secure effective energy management and a coordinated national energy strategy.	Establishes "656" Committee.
NECPA 1978	Promote the use of life-cycle costing in operating Federal buildings, and the use of solar and other renewable energy sources.	Defines Federal Energy Initiative (FEI). Establishes use of life-cycle cost (LCC) method. Establishes publication of Energy Performance Targets. Requires LCC audits and retrofits of Federal buildings by 1990. Establishes Federal Photovoltaic and Solar Programs.
COBRA 1985	Reconcile the budget.	Amends FEI authorizing agencies to use shared energy savings (SES).
FEMIA 1988	Promote efficient use of energy by the Federal Government.	Amends Federal Energy Initiative. Allows Secretary of Energy to set discount rate used in LCC analysis. Removes requirement that agencies perform LCC retrofits by 1990. Establishes energy performance goals for Federal Buildings, 10% by 1995. Directs agencies to establish incentives for energy conservation.

Table 14-1. Federal Energy Management Legislation (*Continued*)

Law	Purpose	Provisions for Federal Energy Management
NDAA for FY1989	Authorizes defense spending.	Establishes incentive for SES contracts in DOD, allowing half of first year savings to be used for welfare, morale, and recreation activities at facility. Other half to be used for additional conservation measures.
NDAA for FY90	Same as above.	Expands DOD's SES incentive to include half of first 5 years' savings.
NDAA for FY 91	Same as above.	Requires Secretary of Defense to develop plan "to achieve maximum cost-effective energy savings"; and simplify SES contracting method. Expands DOD incentives to include utility rebate programs and include two-thirds of savings.

EPCA-Energy Policy and Conservation Act, 1975, Public Law 94-163.
DOEOA-Department of Energy Organization Act, 1977, Public Law 95-91.
NECPA-National Energy Conservation Policy Act, 1978, Public law 95-619.
COBRA-Comprehensive Omnibus Budget Reconciliation Act, 1985, Public Law 99-272.
FEMIA-Federal Energy Management Improvement Act, 1988, public law 100-615.
NDAA-National Defense Authorization Acts: for FY89, Public Law 100-456; for FY90, Public Law 101-189; for FY91, Public Law 101-510.

SOURCE: Office of Technology Assessment, 1991.

called for more simplified SES contracting procedures.

The first Executive Order addressing Federal energy efficiency was issued in 1976 (Executive Order 11912). Through the years, four more orders related to Federal energy efficiency have come down. Table 14-2 summarizes the major provisions of these orders relevant to this discussion.

Table 14-2. Executive Orders Pertaining to Federal Energy Efficiency

Executive	Year	Major Provisions for Energy Efficiency
11912	1976	Delegation of authority; administrative detail
12003	1977	Develop 10-year conservation plan to reduce energy use/sq.ft. by 20% in existing Federal buildings & by 45% in new buildings.
12083	1978	Created Energy Coordinating Committee to assure policy initiatives and resource allocation was coordinated in the major Federal agencies.
12375	1982	Mostly directed at automobile fleet efficiency
12759	1991	Reduce Btu/gross sq.ft. by 20% from 1985 levels; required agency policies to improve industrial process energy efficiency by 20% by 2000; provided for Federal participation in utility demand-side management

The Department of Defense has given each of the four services (Army, Air Force, Navy and Marine Corps) the goal of awarding at least three SES contracts per year, beginning with 1991. Mr. Millard Carr,

Office of the Assistant Secretary of Defense, stressed the opportunities inherent in meeting these goals at a Pentagon Conference in May 1991, commenting that a great potential exists for shared savings contracts. He encouraged those in attendance to take advantage of that potential.

The legislative actions and the Executive Orders have given direction to Federal energy efficiency efforts for more than 17 years. Despite this series of actions promoting cost-effective opportunities to improve energy efficiency within the Federal agencies, much remains to be done.

Related Benefits

The Office of Technology Assessment (OTA) of the U.S. Congress of the United States recently noted that, in addition to the energy cost savings, other benefits would accrue if the Federal government more aggressively pursued energy efficiency opportunities. These potential benefits include:

- promoting the use of energy efficient measures outside the Federal government by demonstrating their cost/benefit and performance;
- accelerating the development and manufacturing of energy efficient technologies;
- learning first-hand which approaches work as a basis for national policy; and
- reducing energy-related environmental and security problems.

OTA also noted that while the benefits are great, consideration must be given to costs, staff allocation and initial capital investments.

Barriers

The reasons usually cited for the inability of Federal agencies to fully implement the use of energy efficient technologies, include:

- Energy efficiency is not central to most agencies' missions. Administrative priorities generally favor needs directly related to the agency's mission.
- Within agency expertise and a shortage of trained personnel reflects the relatively low priority energy holds. This exacerbates

the uncertainty regarding cost and performance of some technologies and approaches.

- A scarcity of capital even to make short term investments, are the results of budget pressures and resource priorities favoring other mission-oriented needs.

- Until recently, there has been a lack of incentives to encourage the agencies to save energy.

- Procedures to assess energy opportunities and acquisition options are generally lacking.

Many of these concerns are not unique to the Federal government. They exist in state and local governments as well as the private sector.

Federal Energy Management Program

The U.S. Department of Energy is the lead agency for federal energy management. The implementation of Congressional and Executive Office directives are managed through the Federal Energy Management Program (FEMP). FEMP is the central mechanism designed to coordinate federal energy-efficiency efforts. FEMP's mission is to exercise leadership in moving the federal government toward greater energy efficiency. Major goals to achieve this mission are to facilitate the transfer of energy management experience and expertise among federal agencies and to improve the efficiency and flexibility of energy use in federal buildings, operations and transportation. Activities include information dissemination, demonstration projects and reports to Congress. FEMP relies on the DOE laboratory system for technical support, especially the Pacific Northwest Laboratory.

With the passage of the Comprehensive Omnibus Budget Reconciliation Act (COBRA) in 1985, which authorized federal agencies to implement shared energy savings, FEMP became active in fostering this alternative financing approach in federal agencies.

Shared Energy Savings

All Federal agencies are allowed to seek private sector financing through shared energy savings (SES) contracts. The SES provisions permit private firms to perform energy services for Federal agencies using their own capital and personnel to improve energy efficiency in an agency. Services include energy audits, acquisition and installation of equipment, modifications to the building shell, operations and maintenance of the equipment and training of Federal personnel. For these services, the energy service company (ESCO) receives a percentage of the savings as specified in the contract. Agencies are permitted to enter into such contracts for up to 25 years.

The Federal government, of course, must pay for these services out of money that would otherwise go to the utility. If the agencies were to do this work themselves, one DOE funded report suggested the savings the agencies could retain by doing the work themselves would be 30 to 70 percent higher. This, of course, assumes the agency would have the funds, would be inclined to dedicate the money to energy conservation work, has the expertise in-house (or could obtain ESCO-equivalent know how) and would pursue savings opportunities aggressively over the same time frame as the contract. Experience in other sectors, including the Federal energy grants program for schools and hospitals, suggests that many of these assumptions are unrealistic.

SES Contract Implementation

Not too long after COBRA was passed, paving the way for shared energy savings contracts, the Congressional Budget Office (CBO) was asked to project the impact of the SES opportunity. Early in 1988, CBO estimated that the Federal government would have 30 SES contracts in place for fiscal year 1988, and these contracts would save the Federal government approximately $250 million over the next five years. As of this writing, however, only eight active SES contracts exist.

The fact that the SES program did not live up to CBO's optimistic projection can be attributed to several factors.

1) Shortly after COBRA was passed, it became evident that Federal employees needed training to deal with unfamiliar aspects of SES, such as changing energy costs, estimating savings, life cycle cost-

ing/energy analysis, rapidly changing technologies, etc.
FEMP in cooperation with the General Services Administration started offering such training in 1989.
Identifying and justifying contracts based on the use of new energy efficient products, technologies and services all based on the nebulous idea of projected energy savings provided a built-in disincentive to change.

2) Many of the firms that offer shared savings contracts were not familiar with Federal procurement procedures.
Many of the barriers built into the Federal contracting process are costly and time consuming for ESCOs. The Federal structure can involve many layers, too many people. Streamlining the process is under advisement, but a few more carrots—and sticks—might help.

3) Some Federal procurement provisions were hard for agencies to implement. OMB Circular A-76, for example, requires that an agency provide comparable data for the contractor costs vis a vis in-house data for a specific service. Such comparisons are very difficult, as the agency may not have comparable talent in-house.

4) Any savings-based contract requires a base year as a reference point from which savings can be calculated. This kind of data was lacking in many instances. A computer program, ASEAM, helps fulfill this need, although ESCOs seem to have had some difficulty in using it.

5) As already noted, the Federal agencies were not allowed to keep any of the savings in the beginning of the SES program and, therefore, lacked internal incentives to pursue these contracts.

6) The short term outlook of some Federal personnel, particularly in the military, causes them to focus on the here and now; not future savings from long term contracts.

7) Extraneous concerns create big obstacles. Unknown plans for sprinkler systems, PCB removal, asbestos, etc. can thwart SES plans.

Questions remain as to whether more incentives would create broader use of SES and whether more simplified contract procedures would prompt more activities by Federal agencies and ESCOs.

SES Contract Experience

Some SES contracts have been in place long enough to give us valuable experiences and a little history. The first SES contract offers some insights.

USPS San Diego Division

In 1987, the United States Postal Service's San Diego Division negotiated the first SES contract to retrofit the lighting system at the Division's San Diego General Mail Facility. The contract removed nearly one thousand fixtures and changed out lamps and ballasts and installed reflectors in 2,292 fixtures. During the initial monitoring, the SES contract exceeded energy savings projections.

Based on the success of the SES light project, surveys were conducted at all the Division's facilities over 3,000 sq.ft. No other SES contracts were initiated in the Division, however, as local utilities provided rebates of 40 percent or more to encourage investment in efficient lamps and electronic ballasts. By using Postal Service investment funds and rebates, the Division is now able to keep all the savings.

The financial aspect of the San Diego Post Office's SES contract created a small problem. They failed to consider the possibility of a decrease in cost per kilowatt hour (kWh). It dropped approximately $0.05/kWh, and a dollar figure had been guaranteed by the contract; so the *dollars* saved were not as good as they might have been.

Figure 14-1 offers a status update on the San Diego Division based on coverage in the second volume of the 1991 Federal Energy Management Activities. The update offers an idea of how a successful project's benefits can spread. The USPS presently has three more requests for proposal for SES services on the street.

Other SES contracts that have been awarded are:

NASA Michoud Assembly Facility, Michoud, Louisiana. May 1988

Corpus Christi Army Depot, Corpus Christie, Texas. September 1988

Naval Hospital, Long Beach, California. September 1989.

Naval Training Center, Great Lakes, Illinois. December 1989.

Naval Air (NAVAIR) Test Center, Patuxent River, Maryland. April 1990.

Fort Shafter, Honolulu, Hawaii February 1991

Oklahoma City General Mail Facility, USPS. December 1991

Lawrence Berkeley Laboratory, U.S. Department of Energy. October 1991

Figure 14-1. USPS Rebate Program at San Diego Division Saves Over a Million kWh per Year

"Utility costs are a bottom line cost that can be controlled; therefore, I expect them to be," says Margaret Sellers, General Manager/Postmaster of the San Diego Division of the United States Postal Service (USPS)—one of the county's largest civilian employers. "We have a tremendous impact on the economy of San Diego; therefore, we must be responsible corporate citizens. It's our duty to be environmentally sensitive and reduce our energy usage by becoming more energy efficient."

Using seed money provided by the USPS National Energy Coordinator, the Division Energy Coordinator, Martin Nelson, targeted lighting retrofits. The result: A project that's already achieved energy savings of over a million kilowatt hours of energy per year—enough to energize 200 homes for a whole year, and saving San Diego Gas and Electric (SDG&E) from having to burn some 1,700 barrels of oil (or 100,000 therms of natural gas) each year.

The San Diego Division of the USPS has a long history of innovation. According to Jim Rankin, SDG&E Senior Account Executive assigned to USPS, "They were the first Federal agency to award a shared energy savings contract. The Federal Government just wasn't set up to handle contracts like these, but their persistence paid off."

The result of this persistence? "We're now more flexible and able to take advantage of many of SDG&E's incentive programs and services, such as their new Lighting Retrofit Program," says Nelson.

Twenty-four postal facilities, with over 60 percent of their utility costs resulting from lighting, have recently undergone lighting retrofits, taking advantage of incentive dollars available through SDG&E's new Lighting Retrofit

Program.

"We've been extremely pleased with the installations," says Nelson. "SDG&E worked with us to ensure that lighting contract worked within our timeframes, and we're appreciative of that," he says. "Out of 24 lighting retrofit completed thus far, we haven't had any employee complaints and we're very happy with the results. The return on investment can't be beat, in fact, we're planning to complete retrofits on 65 postal facilities before we're through," he says.

But that's not all the USPS is doing. The USPS is constructing a 670,000 square-foot general mail facility in Carmel Mountain Ranch and they recently submitted blueprint for the structure to SDG&E for a free review by the utility's New Construction Design Review program. Energy engineers reviewed the plans and made recommendations that, if implemented, will make the new facility more energy efficient.

"Approximately 90 percent of SDG&E's energy efficiency suggestions will be incorporated, including a 600-ton gas air conditioning system, which will save on electricity and, in turn, cost," says Nelson. "SDG&E is providing $50,000 towards the cost of the gas air conditioning system. It will take electrical load off their grid during peak electrical hours, and save us money, because natural gas is less expensive than electricity. It's win-win."

"We've been able to establish a good working relationship with USPS," says SDG&E's Rankin. "The key to these successes lay in the timing, in the creation of relationships with individuals within, who can make a difference and assist with mutual objectives," he says.

"We've been able to do that quite effectively, working together to ensure that SDG&E is aware and involved in any USPS project from the beginning."

According to Jennifer Mitchell, SDG&E's Director of Marketing, "We are so appreciative of the time and effort that has been extended by the USPS/San Diego Division. They have made landmark changes in a very complicated governmental structure, and we appreciate that," she says. "Why, we're even discussing the possibility of converting their fleet to compressed natural gas, a clean-burning fuel that would help USPS to save on uncertain vehicle costs down the road while helping the environment."

"With a fleet of over 2,000 vehicles delivering over 8 million pieces of mail per day to over 1.1 million locations in San Diego County, we've got to be concerned about cost reduction," says Sellers. "Energy and fuel cost represent an opportunity for savings, and it's just good business sense to think ahead and plan for efficiency, particularly when utilities are offering incentives to assist with up-front costs."

"Anywhere we can be environmentally sensitive and reduce our energy usage, we will," say Sellers. And that's the standard operating procedure for the San Diego Division of the USPS.

For more information on the USPS rebate and demand side management program, call Mr. W.G. Eschmann at (202) 268-2596.

Lessons Learned

A few key aspects of SES contracting stand out as we review our progress to date. Lesson 1, and probably the most important, is that an effective SES project absolutely requires a *team* effort. Cooperation and involvement of a number of disciplines in a Federal agency that have a role in SES is key—from the beginning. It's especially important to involve procurement and general counsel from the first.

Lesson 2, the project manager has to pursue the project aggressively and enthusiastically.

Lesson 3, the more we standardize the language and the process, the easier it will become to make use of the advantages SES has to offer. Many provisions, such as termination of convenience, default, cost and pricing language, can be essentially boiler plate. There is movement in this direction. The Navy has a generic solicitation and the Department of Defense has a centralized process.

Lesson 4, as experience with SES expands we have seen, and encouraged, more open-ended solicitations that take advantage of the ESCOs' expertise. The solicitations are becoming more *performance*-based rather than prescriptive closed-ended solicitations. There has also been a move to have a reopen provision where the contractor can come back with upgrades to the equipment.

Finally, there is a move to incorporate the utilities' interests in demand-side management. It is probably the most crucial aspect of performance contracting to come up lately. In fact, the SES solicitation for DOE headquarters requires contractors to apply to Potomac Electric Power Company (PEPCO) for rebates. This involvement can lead to more favorable projects as the rebates leverage the investment. Or, it can reduce the investment level, thereby cutting interest costs, and prompting a quicker payback. As an added plus, the utility's expertise offers Federal back up.

The ground work has been laid. SES contracts have proven themselves as a viable means of reducing energy consumption in Federal agencies. I expect the momentum to pick up. With the requirements on the four branches of the military to do three SES contracts per year, simple math says DOD should have 24 contracts within two years.

All of us involved have learned that to be successful, the SES process has to be a joint undertaking in a non-adversarial "win-win" approach. It needs Federal participants, who sincerely want to reduce energy costs and are ready to enter into a long term enterprise with the private sector to use their expertise and financial resources. That way both parties can accomplish their goals: the contractor receives a proper and reasonable return on investment and the Federal government gains a well-run energy efficient building.

CHAPTER 15

PERFORMANCE CONTRACTING WITH THE FEDERAL GOVERNMENT FROM AN ESCO POINT OF VIEW

Thomas K. Dreessen

THE OPPORTUNITY

For a performance contracting, energy services company (ESCO), the opportunity to implement energy conservation with the federal government (Government) is "mouth-watering" because it represents the largest single energy consumer in the entire United States. It is estimated that the Government consumed $8 billion worth of energy in fiscal year 1990, and of that amount $4 billion was spent on utilities. Of the $4 billion in utilities, approximately 70 percent, or $2.7 billion, was spent by the Department of Defense (DOD) which is comprised of the four armed services; Army, Navy, Air Force and Marines. The energy conservation opportunities are astronomical and could range anywhere from 25percent to 50percent. The magnitude of the opportunity due to its shear size is compounded by the accepted belief that the Government is the largest, single most inefficient and wasteful bureaucracy in the U.S. I believe that a private firm could save the Government another $1 to $2 billion in operation and maintenance (O&M) costs making the total potential savings for this market equal to between $2 billion and $4 billion per year.

The Government has publicly acknowledged that there are significant untapped opportunities for energy efficiency in its buildings, especially as the buildings age. A sense of urgency to cut costs is prevalent due to continued federal budget deficits, which has made energy conservation a priority for both former President Reagan and President Bush. However, in spite of the Government's desire to implement energy conservation on its own, success had been very limited. According to the *Shared Energy Savings Contracting for Federal Agencies,* training manual for federal agencies, the Government has incurred *several* major barriers that have inhibited its ability to effect energy conservation. The most frequently cited barriers are:

- Lack of time and/or technical expertise within the agencies to undertake the needed conservation measures or building improvements.

- The relatively low priority of energy efficiency in many agencies' missions. Events such as the Gramm-Rudman budget reduction process have pushed energy conservation down the list of critical budget items.

- Limited availability of federal funds and the long lag time in obtaining Congressional appropriations after the need for the money has been identified.

- Lack of internal incentives to conserve.

These admitted barriers to implementing conservation on its own has formed the rationale for the Government's entry into shared savings (SES) contracting because Government personnel believe qualified ESCOs have the resources and ability to:

- Identify viable retrofit options and building operating changes to increase energy efficiency;

- Secure project financing;

- Install, operate and/or maintain and repair energy equipment.

SES contracting is defined as a Government contract with a private-sector contractor (hopefully a full-service ESCO) to design, purchase, install, own, finance and maintain specified energy conservation measures (ECMs) in Government facilities, for which the ESCO obtains repayment for providing all equipment and services *solely* from a share of the savings generated from the ECMs. The repayment continues for a specified term, typically from 10 to 25 years. The ESCO puts up all the money required for implementation of the ECMs and assumes total financial risk regarding their performance. If no savings are produced, no payments are made to the ESCO.

In addition to the Government's admission that ESCOs can be a major factor to their achieving energy conservation, the opportunity for ESCOs looks even better when you review the legislative progress effected by the Government to promote SES contracting, as discussed in the previous chapter. The 1980 legislation establishing the Federal Energy Management Program (FEMP) division of the U.S. Department of Energy, the shared energy savings language in the Consolidated Omnibus Budget Reconciliation Act of 1985, Federal Energy Management Improvement Act (P.L. 100-615) in 1988 and Executive Order 12759 signed by President Bush on April 17, 1991 have all encouraged SES as a means of improving energy efficiency in the Government. The Executive Order 12759 essentially provides an opportunity for each agency to directly negotiate with an ESCO, who has been selected by a utility company under a competitive demand-side management (DSM) program.

In specific response to this order, DOD established an implementation plan in May of 1991 to streamline the lengthy procurement procedures and a goal to have each of its four service branches enter into three SES contracts per year through the year 2000 beginning FY 1991. Although the intent was excellent, the initial procedures drafted by DOD did not effect the true intent of the order. DOD only allowed agencies to use the utility selection process as a pre-qualification step to selecting three firms to submit bids under the normal SES bidding process. Since an average of only 2.5 firms have responded to all SES bids to date, this attempt at stream-lining the procurement process renders no value to accelerating SES contracting. Based on testimony by myself and others at a May 8, 1991 meeting set up by DOD to obtain industry comments on their implementation procedures, it appears the procedures will be redrafted to allow the Government to directly negotiate with the ESCO of its choice from the utility's DSM selected list.

In addition to all of the specific legislative action taken by the Government to promote SES, there is a continuing higher level of activity in Congress in response to global warming concerns and a public outcry for environmental improvement in air quality. One of the elements attributed to the deterioration of the earth's ozone level is the use of chlorofluorocarbons (CFCs), which are the primary refrigerant used for air conditioning. The federally mandated change to other refrigerants over the decade of the 90's will significantly reduce the heat transfer efficiency of air conditioners and hence, create demand for more efficient equipment. Other major contributors to poor air quality are SO_2, NO_x and various chemicals distributed into the air from the inefficient burning of coal-fired boilers and other fuel-fired furnaces and industrial processes. The Clean Air Act amendments recently were passed as a result of this public concern. More environmental legislation can be expected, which will enhance the need for energy conservation within the Government facilities.

In summary, the opportunity for SES contracting in the Government is outstanding. The market size, market need and political environment for action could not be better.

The Procurement Process

Now that the opportunity has been defined as "outstanding," it is important to understand how an ESCO implements a SES contract with the Government. The starting point is to first understand the procurement process currently in existence. Although each agency utilizes different procedures, the approach is generally the same as described below.

In the procurement of a SES contract, the Government first defines a specific scope of work to be performed, which is typically included as a minimum requirement for the solicitation issued to the public. In order to identify this scope of work, the Government typically hires FEMP, or an outside consultant, to perform a feasibility study on the designated facilities. An ESCO, or any other contractor contemplating responding to the solicitation, cannot participate in this walk-through or feasibility study prior to the solicitation, or they may subject themselves to disqualification due to having an unfair advantage over the competition.

As you can see, the Government approaches SES contracting like a procurement of a computer, piece of artillery or guided missile. Although somewhat modified by SES legislation, the procurement process follows the Federal Acquisition Regulations (FAR), which were established to identify and specify the items to be procured to the last minute detail and to select the lowest price on a competitive basis. The focus on specifications and competitive prices is what causes the agencies to feel a need to pre-identify the scope of work; so that prices from the ESCOs can be compared on an equal basis.

In some cases, Requests for Qualifications (RFQs) have been issued by certain agencies, ostensibly for the purpose of selecting a short-list of companies to whom a Request for Proposal (RFP) would be issued at a subsequent date. An RFQ requires no pricing or site-specific data; but mostly focuses on the ESCO's prior experience and capabilities to perform the contemplated SES scope of work. In the one instance, where CES/Way participated in a federal RFQ, the short list of ESCOs was selected but subsequently abandoned by the agency because of a fear within the agency that "fair competition" would not exist if the short list was used. The result was that the RFP was issued to anyone who was interested. The issuance of the RFQ was, therefore, a complete waste of time and money for the ESCOs and the Government.

Once the scope of work is defined by the agency, the RFP is advertised in the Commerce Business Daily and a copy of the RFP is issued to any and all parties requesting it. The RFP itself is typically 2 to 3 inches thick and contains virtually all FAR provisions for any type of federal procurement, most of which does not apply to SES contracting. Nevertheless, they must be read and fully understood because the RFP is incorporated into, and in fact, becomes an integral part of the final contract.

A summary of the salient aspects of a typical federal RFP are listed below:

- The agency identifies specific services and required energy conservation measures (ECMs) in the RFP, which the ESCO must include in the scope of work to be paid from savings. In some cases, this includes removal of asbestos and other "non-revenue" producing measures.

- The agency may only allow the ESCO to implement the required ECMs. In other instances, which appears to be the current trend, the ESCO is allowed to offer ECMs in addition to the required ones.

- A standard set of cost and Pricing Schedules must be completed for all ECMs and included in the ESCO's bid. The typical data required includes:
  - — estimated annual energy cost savings for contract term;
  - — estimated ESCO's payments for achieved energy cost savings for contract term;
  - — government's share (%) of cost savings for contract term;
  - — estimated construction costs; and
  - — a *guaranteed* minimum annual energy savings to Government.

- Other standard requirements in the RFP (which are included in the SES contract) specify that the ESCO must:
  - — provide payment and performance bonds on the construction of the ECMs;
  - — agree to the Government's Termination for Convenience (TFC) contract language. (This TFC language allows the Government to terminate its contract with the ESCO at any time and for any reason. Some agencies allow the ESCO to define these termination costs up-front in their RFP responses. However, all contracts, which CES/Way has entered into, only provide for reimbursement of project costs based on an estimate from a third party consultant, who is chosen by the contracting agency);
  - — own the equipment and retain title to it for the term of the contract. The Government will not and cannot enter into any leases for repayment of the ECMs.

The RFP requires the ESCO to provide a firm bid price in its response. If more than one ESCO is initially selected by the Government for further negotiations, each ESCO can revise its price in a subsequent Best and Final Offer (BAFO). However, there is no assurance that a BAFO will be requested; so, the ESCO must be prepared to live with the prices provided in its initial response to the RFP.

In addition to the cost pricing data, the RFP allows the ESCO to provide information on its capabilities to perform the work. The typical data provided are:

- Management approach and organization for executing the work;
- Personnel qualifications and resumes; and
- Previous project experience of similar scope and size.

No weighting of these factors as a criteria for selecting a winning bid has so far been provided in the RFP. The general perception by ESCOs is that the final decision is primarily based on price, which is the net present value of the Government's projected benefit (share of savings) over the proposed contract term to the Government, as "estimated" by the ESCO.

The Operating Process

Once the procurement process is understood, it is important to understand how the SES contracts are implemented and how they operate over the term of the contract.

From an operational standpoint, SES contracting with the Government is very different from other performance contracts. As an example, in the DOD, there is a central group for each armed service which issues the RFPs, selects the ESCO and negotiates the final contract with the ESCO; i.e., the Army's central group is located in Huntsville, Alabama. There is also a contracting officer for the specific armed service who works for the regional operating group of the armed service where the facility is located. Once the contract is signed, the Commanding Officer of the base facility is totally responsible for the contract. The personnel with whom the ESCO negotiated the contract are no longer involved with the project; nor do they have any further authority with respect to it. The base commander/officer assigns management of the contract to one of his/her local procurement officers, who thereafter supervises the implementation and on-going performance of the contract. Unfortunately, the local procurement officer typically has had no prior exposure to the contract at this time, and his/her background is typically centered around the purchasing of standard goods and services for the facility and its personnel under the traditional competitive bid basis.

The Problems

The major problem to implementing any large-scale SES contracting with the Government is that there is no single entity within the Government that has sole responsibility for achieving the mandated 20percent energy reduction. As noted in the above illustration, the commander of each base for DOD has sole authority for making the budgeting and cost reduction decisions within that agency. Although FEMP has been charged with the task of promoting SES contracts to the various agencies and locations, it has no authority to force the agencies to enter into them. Nor, in fact, are the agencies and thousands of base commanders, or other facility managers, in any way required to work with FEMP in procuring the SES contracts.

The individual agencies have little incentive for entering into SES contracts. Although DOE gives annual Federal Energy Efficiency awards to recognize the achievement of federal energy managers, the availability of additional incentives for other agency personnel is virtually nonexistent. The federal personnel in general are not interested in pursuing SES contracts because, if they fail, they risk the loss of their job; if it succeeds, they may get a "pat on the back." When you combine this risk/reward relationship with the inherent resistance to change in a huge public bureaucracy like the federal government, the likelihood of having any major success is remote. It is much easier for the established bureaucrats to stay within their existing procurement and administrative procedures, which require no changes and allows them to retain control of their facilities and their respective utility budgets if SES contracting is not pursued.

Resistance by several of the agencies to include avoided maintenance in the calculation of savings achieved from the installed ECMs is another impediment. The existing legislation primarily addresses energy savings; therefore, the agencies, in most cases, have not allowed maintenance savings to be included in any savings calculations. On an old, high-maintenance facility, the inclusion of maintenance savings is instrumental in allowing an SES approach to work. The taxpayers would also benefit more from the capital improvement because it reduces not only energy, but also wasted maintenance costs.

A discouraging trend from the ESCOs' perspective, which is happening within several of the agencies, in particular the DOD, is the implementation of quick payback items on their own, in lieu of pursuing SES contracting. The Navy has recently implemented numerous Operation and Maintenance of Energy Services (OMES) programs for many of its facilities. These programs entail the Navy hiring a maintenance-type contractor to replace/repair equipment such as steam traps, valves and other maintenance items which payback in 2 years or less. The Navy pays for them out of existing budgeted funds.

Another discouraging trend is that the DOD is encouraging large Army and Naval bases to negotiate directly with their respective utilities to obtain cash rebates from the utility DSM programs. The rebates become available by installing more efficient electric lights and motors. The equipment is typically installed by lighting contractors (not ESCOs) chosen by the utility company. Although the utility provides a cash benefit subsidizing the payment of these measures, the effect of installing just these measures promotes a "cream-skimming" approach, which negates a future opportunity for many of these facilities to implement full-scope energy conservation retrofits under a SES contracting. On the near term, it would appear that the Government is achieving energy cost savings in the most cost-effective manner by instituting quick payback or utility rebate items. This ignores the concept of Total Energy Management put forth by the National Electrical Manufacturers Association and others in the mid-1970s, and accepted ever since by effective energy managers across the country. By looking to the single measure approach, an agency is doing "spot remedies" and ignoring the greater savings that could be achieved with a comprehensive approach. The quick turn around is seldom the most cost-effective in the long run. One DOE study, for example, placed lighting fourth in ranking cost-effectiveness—behind controls, heating and ventilation measures. The loss of these long term savings, in nearly every case, will exceed any rebates received. While the intent is good, the results is that long-term energy reduction will be limited, which will make it very difficult, if not impossible, for the 20percent reduction goal to be accomplished. The savings generated from the quick payback items, such as lights and motors, are essential to any full-scope energy/maintenance reduction program. The measures, in effect, subsidize the long payback capital improvement items that generate the long-term energy and maintenance benefits to the facilities. An ESCO cannot only access the utility rebates and incorporate them into a full-scope SES program, but also has the ability to leverage the savings paid for by the utility and significantly increase the magnitude of non-facility capital improvements that can be done. If done under the auspices of a SES contract, the savings are retained in the agency's budget to pay the ESCO; if done by the agency, the savings are lost in next year's utility budget.

The fact that only eight SES contracts have been entered into with the Government since SES inception in 1986, indicates the magnitude of the problem with the procurement process. The major contributing factor to these poor results is the degree to which the procurement/ contract process is encumbered by existing FAR regulations. These regulations are extremely complex and do not apply to SES services. FAR regulations were written to obtain specific goods and services which have a fixed price and pre-defined specifications. Under SES contracting, the fixed price represents a net present value calculation of the estimated savings from the ECMs implemented over a 10- to 25-year period of time and in no way reflects just an equipment bid. The regulations also do not apply since a private company owns the equipment for the life of the project and makes the equipment available for use for a long period of time. Additionally, the exact payment amount for this equipment is unknown because it varies with the actual savings. This is foreign to any other procurement process currently used by the Government. The attempt by various agencies to implement SES contracts under existing FARs has resulted in a very complex 2- to 3-year procurement process and a general lack of interest from the federal agencies and the ESCO industry. Legislation before Congress to modify FAR with respect to SES is critically needed.

The RFP selection process places entirely too much emphasis on price versus the qualifications and prior experience of the ESCO submitting the response. As stated above, it is very difficult to compare the responses from two companies when the price provided in each one's response is only as good as the savings estimates and assumptions behind them. The only way to determine the value of a price offered by an ESCO is to look at the track record of the ESCO. The degree to which actual savings achieved relate to the estimated savings provided in the ESCO's original response to a customer is the only reliable gauge.

Another major problem is the extraordinarily high cost the ESCO must absorb in order to provide a response to an RFP. Although there are some recent attempts to reduce this cost, not all agencies have adopted any change to the typical process. The primary cause for the high cost is the requirement that the ESCO provide a firm bid price in its response to the RFP. In order to provide this, an ESCO must perform a complete facility audit to enable it to accurately estimate the savings and costs to a point where an ESCO can live with them for 10 to 25 years. In some instances, a short list is created, which provides the qualified ESCOs a higher probability of being selected, and helps justify the risk of incurring the high audit cost.

Several other items increase the cost of doing business with the Government. They center around the Government's inability to enter into any leases for the equipment, which precludes an ESCO from financing the project with fixed lease payments over time. The Government's inability to lease equipment results in the ESCO incurring very high financing costs and assuming a significant larger amount of risk in implementing the ECMs. These costs, of course, increase the cost to the Government. In effect, the Government ends up saving less energy for its level of investment.

Other major elements which create high project costs is the Government's requirement for TFC language; and, in some cases, a requirement that the ESCO provide a guarantee of savings. These last two requirements drive up the cost of doing business with the Government far beyond any other market sector in the country. It is particularly unreasonable for the Government to require an ESCO to guarantee a savings level when the ESCO is putting up all the money and only gets paid on the savings actually achieved. All of the financial risk rests with the ESCO. The additional risk of having to provide a minimum return to the Government, even if the savings are not sufficient to make a debt service payment to the ESCO, requires the ESCO to place a substantial premium return on the project cost to cover this risk. The same premium return is added because of the TFC language.

A final problem in the procurement process is that the RFPs may not only require the ESCO to install pre-defined ECMs (which may or may not have good paybacks), but may not allow the ESCO to add any additional measures to the project. This clearly eliminates the creative resource capabilities of the ESCOs from which the Government could benefit, for it does not allow the ESCO to maximize the capital improvements and operating cost reductions in a facility. In cases where ECMs are limited to lighting, although payback are good, the facility may be limited from any future inclusion in a full scope retrofit as previously described.

After the contract is signed, a separate set of problems exist, which must be corrected in order to keep ESCOs interested in performing SES projects for the Government. The major problem is the lack of continuity. Once the contract is signed, the regional people who selected the ESCO and negotiated the contract are no longer involved, nor do they have any authority in the implementation phase of the contract. More importantly, in some cases, the local contract officer, who now is in charge, has never even seen a copy of the contract, has no prior knowledge nor experience in SES contracting and, therefore, typically has an adversarial attitude toward the ESCO. Our company alone has suffered substantial losses because of this organizational flaw. To illustrate the potential magnitude of this problem, on one of CES/Way's projects, old inefficient chillers were replaced with new efficient ones. CES/Way was to receive its SES payments based on the amount of chilled water produced. Although we diligently tried to establish a minimum load of chilled water production in our contract negotiations, it was prohibited by the agency (presumably FAR). However, we were assured by the regional contract negotiators that the likelihood of any reduction in chilled water below historical levels was highly remote and that if it occurred, we could obtain a contract adjustment. Subsequent to implementation, a local employee (who had no understanding of our contractual terms) decided he could save a lot of energy by turning off the chillers and eliminating air conditioning at the facilities. He continued this for several months until the occupants finally complained about the hot working conditions from having no air conditioning. Despite our continual appeal for adjustment, nothing was done.

There are so many layers of approval in the Government, that it takes nearly 6 months to 1 year (or more) to get changes to a contract consummated. This is caused by a lack of priority plus a lack of knowledge on how to deal with SES contract changes throughout the plethora of administrative levels, each of which must approve the change.

THE SOLUTIONS

I believe the single most important solution required for the federal government to effect SES contracting to a point where it will significantly contribute to the Executive Order's 20% reduction by the year 2000 is to establish a Central Agency which will issue all RFPs, review all responses, make ESCO selection recommendations to the agencies, negotiate final contract terms and resolve all contract disputes throughout the term of the Contract. With this approach, the local personnel at each agency would have to be included in the entire process, from beginning to end. The local agency personnel would be responsible for making the final ESCO selection and for approving all site-specific decisions related to the contract. A master contract should be developed for use by all agencies, which would have all general terms included. The site specific data would be attached as schedules with the technically oriented items (such as baseline methodology) to be negotiated by the Central Agency and the other site-specific decisions (type and location of equipment installed) to be negotiated by the local agency.

The Central Agency would have to be comprised of at least one technically competent individual from each agency. The Central Agency would have the sole responsibility and authority to achieve the 20percent energy reduction by the year 2000, and it would also have the authority to prioritize and select the facilities who would issue the RFPs.

The procurement regulations should be rewritten and standardized to follow the selection procedures used by many public sector institutions across the country in their selection of architects and engineers. The conflicting aspects of FARs should be eliminated. The new procurement method should entail having the Central Agency issue an RFQ which requires any interested ESCOs to submit a response documenting its:

- prior experience in implementing performance-based projects;
- actual savings results and original savings estimates on prior projects;
- size and scope of prior projects;
- personnel and operating procedures used for implementing projects; and

- financial and bonding capabilities.

Oral interviews would be conducted by the Central Agency to determine which ESCO would be put on a list to be eligible for all future SES contracting RFPs. All these pre-qualified ESCOs would be eligible to respond to the RFPs. Any firm could be put on the list at any time, subject to meeting the qualification criteria.

Once an RFP was issued by the Central Agency, its procedures would follow a standard form and include the following changes from existing SES procurement procedures:

- Maintenance and any other documented operating savings could be included in savings calculations;

- Prices included in an RFP would be subject to final audit with a 10percent tolerance band resulting in a requirement that final savings achieve at least 90percent of the estimate provided by the selected ESCO in its RFP response. If 90percent is not achieved, the agency could select the next candidate at no cost to the agency;

- The criteria for selection would be weighted and would emphasize the experience and qualification at a minimum 50% level;

- Leasing would be allowed to the extent that an ESCO could satisfactorily guarantee to the Government that savings would minimally equal or exceed the lease payments;

- Specific pre-agreed TFC prices would be provided in the ESCO's response (DOD is already doing this in its new procedures); and

- Although agencies may pre-define measures in the RFP, the ESCO would be allowed to include additional measures and alternatives in its response.

To prevent the exclusion of those facilities which had already implemented quick payback programs such as OMES, utility rebate or other non-SES programs, the net savings to the Government (assuming a minimum 10-year amortization of the installation cost paid by the

Government) would be allowed to be credited to the savings calculation for any subsequent SES contract.

In summary, I truly believe these solutions, if implemented, could have a profound positive impact on SES contracting.

The Future

The long-term performance-based SES contracting opportunities for the Government facilities are enormous. The operative word to emphasize "long-term" because of all the existing complex and archaic procurement regulations and other bureaucratic barriers that have to be overcome. The good news is that FEMP, DOD and other agencies have many competent and conscientious employees who are very optimistic and enthusiastic about implementing SES contracting. They are "handcuffed," however, by the regulations and lack of authority to get the job done.

Although many who read this chapter may say my suggested solutions are unrealistic and can never be implemented; to them I would suggest that these solutions are no less realistic than the goal of achieving 20percent energy reduction in the federal facilities by the year 2000 under the current regulations in place today.

SECTION 2.
THE UTILITIES

Federal and state regulatory agencies have been steadily increasing the pressure on utilities to do least cost planning. The Clean Air Act Amendments have made investments in energy efficiency more attractive; and in some instances, imperative. Regulators and utilities have come to generally accept that an investment in conservation should be part of a least cost strategy and an environmentally responsible posture for utilities.

For years, utilities have offered technical advice and assistance to aid customers' energy efficiency efforts. Utility rebates, time of use rate schedules, curtailment programs have all given utilities a way to manage consumption and loads, as an alternative to purchasing supplies or creating more generating capacity.

As conservation has become an option to additional supply, more attention has been focused on the *demand* customers place on a utility at any given point in time. This demand is what really tests a utility's capacity, and ultimately forces the purchase of supply or more generating capacity. Demand-side management (DSM) has the potential to offset such supply meeds; so it has become a focal point in least cost management.

Under the provisions of the Public Utility Regulatory Policy Act (PURPA), utilities have purchased supply from independent power producers (IPPs) on a competitive basis. Gradually, it has occurred to regulators and utilities that the opportunity exists to purchase conservation and DSM in much the same way. One of the more intriguing aspects of this concept is that the bidding process gives substance and documentation to a least cost criterion and also quantifies the availability and price of bulk purchase energy efficient resources.

Providentially, the performance contracting industry was in place and ready to assist utilities manage the demand-side of their business—and to do what the utilities have not been able to do for themselves; offer *guaranteed* results. It would seem to be an ideal marriage of needs and capabilities. But the courting process has not been smooth. For example, verification of savings, particularly for utilities, has been a bone of contention; and regulators are beginning to require documentation.

In Chapters 16 and 17, Mr. David Wolcott and Mr. Cary Bullock offer special insights into these concerns. Their observations may help all parties understand and appreciate the perspective and needs of others involved in this seemingly simple, but unfortunately complex, endeavor.

CHAPTER 16

ENERGY PERFORMANCE CONTRACTING AND DEMAND-SIDE MANAGEMENT

David R. Wolcott

The fundamental goal of energy performance contracting is the achievement of energy savings. In the world of electric and gas utilities, achieving energy savings is called "demand-side management."

Introduction to Demand-Side Management

Demand-side management (DSM) is a phrase that reflects utility and regulatory recognition of the resource represented by a utility customer's electricity usage. Historically, utilities have met customer demand for electricity through "supply-side management" activities, such as building new utility power plants. In the last decade, options have increased to include acquisition of power from non-utility power producers.

Utilities and regulators have come to understand that electric resources also exist on the customer's side of the meter and it may be as cost-effective to reduce electric demand as to build expensive new power plants. A "negawatt" obtained through DSM is just as good as a megawatt of generation capacity. DSM is achieved through *energy con-*

servation; a reduction in the consumption of kilowatt hours (kWh) or heating fuels, such as natural gas (mcf) or oil (gallons), or *load management* which is the reduction of kilowatts (kW) of power demanded or the displacement of that demand to off-peak times.

Technical Potential of Demand-Side Management

To understand the opportunities for DSM, we first should gauge the currently available energy efficient technology. Then, narrow our focus to that fraction of the technical potential that is cost-effectively achievable through DSM programs. An understanding of the technical and achievable potential is necessary to estimate the possible impact of DSM as an electric resource (Wolcott, 1991).

The New York State Energy Research and Development Authority engaged the American Council for an Energy-Efficient Economy (ACEEE) in a 3-year project to analyze the technical and achievable DSM potential in New York. The project was cosponsored by the New York State Energy Office and the Niagara Mohawk Power Corporation. The first analysis focused on technical potential and developed the "conservation supply curves" which have become a standard for comparable analyses in other states (ACEEE, 1989). Figure 16-1 is ACEEE's conservation supply curve for New York, based on installed capital costs, each horizontal line representing the amount of energy savings available from a particular DSM measure.

ACEEE found that the technical potential for DSM is 50 percent of current consumption in New York's commercial sector, 37 percent in the residential sector, and 22 percent in the industrial sector. Full adoption of all the DSM measures analyzed in the study would reduce New York's aggregate electricity consumption by 34 percent. This result is generally consistent with a range of national estimates of DSM potential bounded by the Electric Power Research Institute (22-42 percent) at one extreme and the Rocky Mountain Institute (75 percent) at the other extreme (Fickett et al., 1990).

Achievable Potential of Demand-Side Management

ACEEE applied the 21 most successful DSM programs from across the nation to the building stock and electricity end-uses of the three largest utilities in New York (Long Island Lighting Company, Consoli-

Figure 16-1 Electricity Conservation Supply Curve for New York State

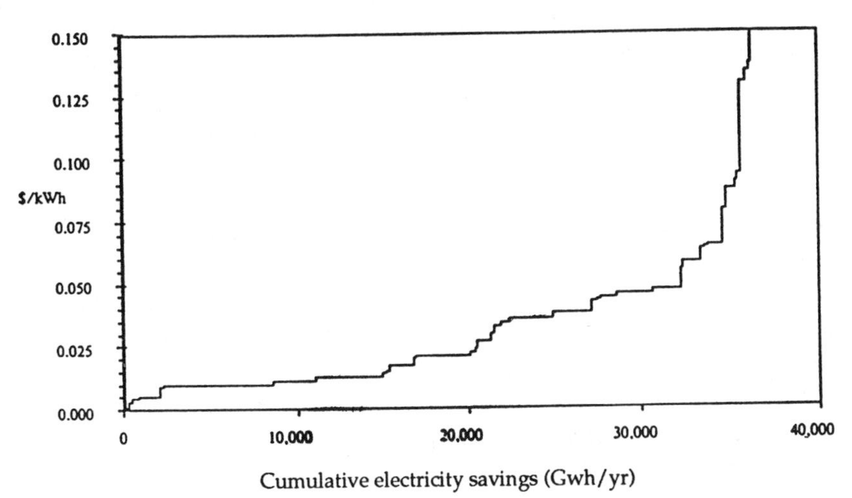

Source: ACEEE, 1989

dated Edison Company, and Niagara Mohawk Power Corporation, comprising 80 percent of electricity use in New York) [ACEEE, 1990 (2)]. The results of the study indicated that aggressive, cost-effective DSM programs for the three utilities would cost over $320 million per year, would raise retail rates between 2-9 percent, and would meet over 50 percent of new load growth for those utilities to the year 2008. This would mean that six 600 MW power plants would not have to be built.

The study also concluded that 80 percent of the technical DSM potential can be achieved cost-effectively, of which roughly half can be accomplished through utility programs and the other half through market responses to price, building codes, and equipment efficiency standards.

Integrated Resource Planning

The concept of cost-effectiveness is key in gauging the size and viability of the DSM market. To see the importance of this concept, we must understand the significance of the fundamental regulatory reform initiative known as "Integrated Resource Planning (IRP)." IRP is also called "least-cost" planning because it refers to the concept of utilities acquiring electric resources at the lowest possible cost. In this sense, an "electric resource" is the next increment of capacity that a utility must deploy in order to meet its customers' energy and power requirements. Resource options typically include construction of new power plants, purchase of electricity from other utilities or independent power producers, and achievement of conservation that frees up existing capacity.

When IRP first became fashionable, most utilities were offended! Least-cost planning could be interpreted as a judgment that utilities had *not* been acquiring the lowest cost resources. In fact, most utilities had been acquiring the lowest cost resources all along, however, those decisions had been based on a cost-effectiveness analysis that considered the benefits and costs of competing investments strictly from the utility's point of view. This observation highlights the significance of the new cost-effectiveness tests which have been introduced through IRP: the *Total Resource cost (TRC) Test* and the *Societal Test*. Following is a discussion of the original *Utility Test* and the two new tests in terms of their significance to the development of the DSM market (CA, 1987).

Utility Test

A benefit/cost (B/C) analysis involves a comparison of the benefits and costs of an enterprise which is considered cost-effective if the ratio of benefits to costs exceeds one. In the old days, when a utility looked at a DSM resource, a simplistic B/C tally looked like this:

BENEFITS	COSTS
Avoided Operating Costs	Program Administration
Avoided Capital Costs	Subsidies
	Lost Revenues

On the benefits side, "Avoided Operating Costs" are derived from the fuel and operations and maintenance savings that the utility realizes by not having to generate power because of reduced consumption from conservation.

"Avoided Capital Costs" refer to the new capacity that the utility does not have to build to meet its customers' peak demand requirements. On the costs side, "Program Administration" refers to all of the utility's direct costs to design, implement, and evaluate the DSM program.

"Subsidies" are incentives such as a rebate which the utility makes to customers and/or vendors to motivate them to participate in, and carry out, the program. "Lost Revenues" are the sales the utility forgoes when a customer does not buy electricity because conservation has reduced its demand.

Using the Utility Test above, the most cost-effective DSM programs are those that achieve load management: reducing a customer's peak demand requirements (peak clipping), helping the utility avoid the need to build new capacity. The best load management measures shift the customer's demand from a peak period to a non-peak period (valley filling), such that the cost of revenue losses is eliminated at the same time that the benefit of avoided capital costs is realized. However, cost-effective load management measures constitute only about 25 percent of the achievable potential available from utility DSM programs. Therefore, if DSM programs had to pass the Utility Test, the utility DSM market would be one-fourth its potential size.

Total Resource Cost (TRC) Test

A major innovation under IRP that has received widespread adoption is the consideration of benefits and costs of DSM programs from both the utility's *and* the customer's point of view. The TRC Test does this by adding a new item to each side of the B/C tally:

BENEFITS	COSTS
Avoided Operating Costs	Program Administration
Avoided Capital Costs	Subsidies
Customer Bill Savings	Lost Revenues
	Customer Costs

On the benefits side, "Customer Bill Savings" are the savings over the useful lifetime of the DSM measure that the customer does not have to pay the utility. On the costs side, "Customer Costs" refer to the portion of the total installed cost of the DSM measure that is not paid for by the utility through subsidies. Many more DSM measures become cost-effective by adding customer bill savings to the benefits side of the tally. Doing so increases the cost-effective market for DSM from 25 percent (load management alone) to 100 percent of the achievable potential for utility-sponsored DSM.

Societal Test

Another innovation under IRP that is under consideration in many jurisdictions is the consideration of the costs and benefits of various electric resource options from society's point of view. Under economic theory, such a step introduces "externalities" into the equation: those costs and benefits that are not reflected in electric rates. Most utility externalities are costs of environmental damage that results from pollution emitted from electric generators (even after they have complied with all applicable environmental regulations).

DSM measures are generally considered to be environmentally benign. Therefore, the B/C tally under the Societal Test includes a credit for the "Avoided Externality Costs" of pollution that was not created through electrical generation as a result of reduced consumption from conservation, as indicated below:

BENEFITS	COSTS
Avoided Operating Costs	Program Administration
Avoided Capital Costs	Subsidies
Customer Bill Savings	Lost Revenues
Avoided Externality Costs	Customer Costs

To give an indication of the magnitude of this benefit, in New York the avoided externality benefit of DSM has been valued at 1.4 cents/kWh. In other states; e.g., Massachusetts and Nevada, the amount is even higher, depending on the electricity generating combustion technology that is displaced by DSM. By adding this value to the benefits side, the amount of DSM that is cost-effective is greater than 100 percent of the achievable potential based on the TRC Test.

Regulatory Incentives for Demand-Side Management

A major disincentive to enthusiastic utility implementation of DSM has been the simple fact that utilities lose money when their customers use less electricity. Under traditional regulation, every kWh that a utility sells will increase its profits and every kWh that a customer *does not* use decreases utility profits. Since the profits earned by investor-owned utilities largely benefit shareholders, it's no surprise that utilities have viewed DSM as a threat to their financial viability.

An important component of the development of IRP has been the recognition of this problem and the development of "regulatory incentive" mechanisms that remove disincentives for utilities undertaking DSM, and in some cases, provide positive incentives (Moskovitz, 1989). The main utility concern is the recovery through rates of all of the costs of program administration, subsidies, and lost revenues. If a utility can allocate all these costs as revenue requirements, it will be indifferent to the effects of DSM on its balance sheet. Another way that a utility can be indifferent to DSM is if its profits are "decoupled" from sales. Historically, utilities earned an allowed rate of return that was collected through revenues. If sales went down, so did profits. An important regulatory incentive that was implemented in California during the 1980s, the Electric Rate Adjustment Mechanism (ERAM), set utility profits independent of sales which made utilities indifferent to the fact that their customers used less electricity because of their participation in DSM programs.

Since then, ERAM-type mechanisms have been implemented in a number of other states including New York, Maine and Washington (Nadel et al., 1992). Other regulatory incentive mechanisms have been developed to give utilities extra profits for successfully implementing their DSM programs in addition to removing the disincentives. There have been many variations on the theme with the most popular one giving utilities a percentage share of the total resource savings (say, 20 percent of the difference between total savings minus total costs as defined in the B/C lists above). Other approaches have included the provision of a "bounty" or pre-specified amount per unit of energy saved or an additional amount added to a utility's allowed rate of return on either DSM expenditures or its entire rate base.

Cost recovery and decoupling mechanisms do not change the B/C analysis previously described. However, the provision of positive incentives does add an additional item to the cost side of the equation. While this additional cost reduces those DSM measures that could be considered cost-effective, most agree that this would be a small price to pay to gain a utility's enthusiastic implementation of aggressive DSM programs. As such, regulatory incentives are generally recognized to be a key component of any IRP program.

Status of Integrated Resource Planning

In 1990, the Edison Electric Institute (EEI), found that IRP was being practiced in 23 states and was in the process of being implemented in 8 more (EEI, 1990). From a more recently published report, EEI estimated that regulatory incentives for DSM are in place or under active consideration in 30 states (Barakat, 1991). Lawrence Berkeley Laboratory (LBL) conducted a survey which revealed that in 1990, 12 states had incorporated environmental externalities in their IRP process while 5 more were actively considering doing so (LBL, 1990). In April 1991, LBL completed a survey which found that the IRP concepts originally developed for electric utilities were being extended to natural gas regulation in 15 states [LBL, 1991 (1)].

A general observation is that 15-20 states are involved in all these activities including many in the northeast, the upper midwest, and the west coast. These high population areas constitute the core markets for DSM. The evolution of IRP is a dynamic process. Whether it survives and flourishes or dies as a regulatory concept depends on a number of influences that are currently either threatening or promoting IRP.

Influences that Threaten IRP

IRP and DSM are under attack by two groups: large industrial customers and the "trades" (e.g., electrical, mechanical, and plumbing contractors) (Anderson et al., 1991). The large industrial customers are represented nationally by the Electric Consumers Resource Council and are represented in various states by other groups; e.g., Multiple Interveners in New York. The industrials fear cross-class subsidization, as they argue that they will unfairly bear the costs of all the utility DSM programs. This argument is based on the assertion that they have already done all the cost-effective DSM (by responding to prices going back to the original energy shocks of 1973) while the residential and commercial customer classes have not. This assertion is generally not borne out by the analysis of DSM potential in industrial facilities that was done in New York (ACEEE, 1989).

The industrials are also offended by what they view as regulatory arrogance in the incorporation of environmental externalities. Their argument is that society has established a politically acceptable level of pollution control through laws; e.g., the new Clean Air Act amendments.

They say that efforts by state regulatory authorities to exceed that level by considering externalities in utility planning is an inappropriate expansion of their jurisdiction.

The issue with the trades is unfair competition. They see utilities directly involved in performing DSM installations, often through their own unregulated subsidiaries, as having an unfair advantage in cost recovery and allowed rate of return. They say that utility DSM programs are forcing them out of their traditional markets. They also say that utility programs have expanded too far too fast and have outstripped the technical capabilities of the workforce.

Influences that Promote IRP

There are mitigating factors at play that address each of the threats to IRP mentioned above. For example, the Association of Demand-Side Management Professionals, a recently established professional association made up largely of utility personnel and consultants, has initiated a serious effort to work with utilities and their industrial customers to design DSM programs that provide real and direct benefits to the industrials, so they don't feel that all they see are DSM program costs. Such programs typically provide custom-designed incentives for efficiency improvements in industrial end-uses that often don't fit into the neat categories of the utilities' conventional rebate programs.

Another development that will blunt the industrials' attacks is the large number of legislative proposals on Capitol Hill and in the states that would provide statutory authority to many of the components of IRP. For example, versions of the National Energy Strategy Act debated in Congress in 1991 included provisions for establishing regulatory incentives throughout the United States. Five bills introduced before the New York legislature in 1991 would codify IRP, regulatory incentives, and the incorporation of environmental externalities into law. The new Clean Air Act Amendments provide special pollution credits to utilities that engage in an IRP process and aggressively implement DSM.

As for the concerns of the trades, the introduction of DSM bidding programs provides a mechanism for utilities to hire vendors to deliver DSM services directly to their customers. There is no competition with the trades; the utility programs are opening up opportunities that would otherwise not have been available to them. For example, most participants in DSM bidding programs have been energy service companies

(ESCOs) that act as general contractors, often hiring the trades as subcontractors.

Another notable development is the recent establishment of the Demand-Side Management Training Institute which has been organized to coordinate the delivery of training services on DSM and IRP issues to utilities, regulatory staff, utility customers and other participants in the DSM industry. Such activity will ensure that the regulatory momentum behind IRP will not get out in front of the capabilities of the industry to deliver the goods.

Demand-Side Bidding

One of the tenets of IRP is that all energy resources are evaluated on a "level playing field"; i.e., no particular resource has an inherent advantage over another. One of the approaches that has evolved to implement this strategy is the acquisition of new resources through competitive bidding, including demand-side bidding. This development is significant because most all performance contracting in utility DSM programs has been undertaken by ESCOs participating in demand-side bidding programs. In these programs, the utility first enters into a contract with an ESCO that agrees to deliver a certain quantity of DSM for a payment at a specified price. The ESCO then markets its services to the utility's customers and enters into separate performance contracts with them.

Behind the concept of bidding is the idea that before a utility considers building a new power plant, it must first seek lower cost sources of power whether from independent power supplies or DSM. Non-utility entities are invited to competitively "bid" a price that the utility will pay to acquire these lower cost resources. Some view such a payment as a financial incentive or a subsidy when, in fact, it is the utility's cost of acquiring a resource for less than its incremental cost of providing that resource itself.

"Competitive bidding" is actually a misnomer as it implies a selection decision based solely on the price bid, as in a public procurement process. However, all bidding programs to date have also used, to varying degrees, non-price selection criteria such as:

- Technologies to be installed (Do they meet the utility customer's needs?);

- Marketing approach (Will the program be successfully implemented?);
- Experience and qualifications of bidders (Have they undertaken this type of program before?);
- Financing approach (Does the project have the necessary backing?);
- Environmental impacts (Are they minimized?); and
- Security (Is the utility protected from nonperformance?).

DSM bidding is an auction in which a utility solicits proposals from entities interested in achieving specified amounts of DSM savings in its service territory; e.g., 1,000 kW of demand reduction. The proposals are evaluated in terms of criteria such as those mentioned above and program participants are selected competitively. The utility then pays the price bid; e.g., $500/kW, for DSM savings that are estimated or measured. If the bidder fails to deliver the promised amount of DSM savings on time (typically within two to three years), it forfeits a security deposit it has posted.

There are many variations on the theme. Eligible bidders can include ESCOs that develop projects on a performance contracting or "shared savings" basis, other vendors, or the utility's own customers. Bids can be structured as the price to supply a block of kW demand reductions, kWh energy savings, or both. DSM bidding can be undertaken in a fully integrated program with supply-side bidding to acquire independent power resources, or as a stand-alone program. The utility can target certain end-use sectors; e.g., commercial and industrial, or allow an all-inclusive program. Payments to bidders can be made once or over time in installments. These variations frame many of the strategic program design and evaluation issues of DSM bidding.

Evolution of DSM Bidding

DSM bidding is currently being implemented by some utilities as part of integrated, all-source bidding programs in which DSM and independent power supply-side resources are acquired through a single solicitation. This development in DSM bidding can be traced historically from two trends in the utility industry during the 1980s: supply-side bidding auctions for independent power and customer incentives for DSM.

The Public Utility Regulatory Policies Act (PURPA), opened up the market for non-utility independent power producers in many parts of the United States. PURPA guaranteed independents nondiscriminatory access to the electric grid and payments for their power set at the utility's avoided cost of producing or purchasing the power itself.

However, some utilities, with the approval of their regulators, sought to control the development of the most cost-effective independent power by conducting auctions. The first program, implemented by Central Maine Power (CMP) in 1983, established a supply block that independents were invited to compete to fill by bidding prices against each other. A ceiling price for bids was set at CMP's avoided cost. The utility proceeded over the years to offer auctions for "decrements" of supply reflecting CMP's decreasing avoided costs as previous supply blocks were filled. Supply-side bidding programs for independent power have subsequently been implemented by utilities in 32 states (NIEP, 1990).

Customer Incentives for DSM

Some of the first utility DSM programs implemented in the early 1980s involved the payment of a financial incentive by the utility to induce a customer to buy an energy-efficient product. Such rebate programs for energy-efficient appliances and lighting are now the core of most utility DSM programs. While a rebate can be viewed strictly as an incentive payment by the utility to motivate particular customer behavior, it can also be viewed as a proxy payment by the utility for the DSM resource that will be provided by the customer's installation of the energy-efficient measure; thus, rebates constitute implicit utility payments for DSM resources.

The first case of an explicit utility payment for DSM resources was the Bonneville Power Administration's "Purchase of Energy Savings" program, field tested in 1983 and piloted in 1985. In that program, ESCOs responded to a Request for Proposals (RFP) and were selected based on their qualifications. The ESCOs provided audit, installation and financing services and were paid for energy saved up to the avoided cost ($.05/

kWh) for the life of the measures installed (maximum 12 years).

Utilities regulated by the Massachusetts Department of Public Utilities were the first to combine the "pay for savings" concept with a bidding process through which ESCOs were selected based on qualifications as well as the amount of savings bid. Pilot programs implemented in 1986-87 included Boston Edison's Encore program, New England Electric Systems' Performance Contracting program, and Commonwealth Electric's Guaranteed Savings program. Each program presented variations on the theme, but basically ESCOs provided audit, installation, and financing services and were paid for kWh savings up to avoided cost or baseload kW saved up to a ceiling price (Cole et al., 1988).

Integrated All-Source Bidding

The first utility to integrate the price bidding of independent power-supply side projects with the DSM "pay for savings" concept was Central Maine Power. In 1987, as it was running one of its supply-side auctions, CMP received a request for participation from an ESCO desiring to bid a DSM resource. Using that bid as a model, CMP gained approval from the Maine Public Utilities Commission to implement its Power Partners program on a pilot basis allowing both customers and ESCOs to submit bids for either supply or DSM resources. Supply and DSM bids were evaluated side-by-side with the same criteria and were selected based largely on bid amount and price. The program expanded with the issuance of an RFP in 1989.

The integrated all-source bidding concept saw full-scale adoption by the New York Public Service Commission in June 1988 (NYPSC, 1988) and the New Jersey Board of Public Utilities in August 1988 (NJBPU, 1988). Utilities in both states started implementing programs in 1989; e.g., Orange & Rockland Utilities (ORU) and Jersey Central Power & Light. However, the New York utilities were given the option of running integrated supply and DSM bidding programs (such as Niagara Mohawk Power Corporation did) or separate, parallel programs (such as ORU did). All New York and New Jersey utilities had DSM bidding programs underway by 1991. Implementation of DSM bidding continues to spread. At last count, 14 utilities have implemented DSM bidding programs, including utilities in Washington and Indiana in addition to those mentioned above (Goldman and Busch, 1991).

An interesting development in the ESCO industry has been the utility response to the business opportunities promised through DSM bidding. A number of the ESCOs responding to utility solicitations are themselves the subsidiaries of utilities. For example, the first ESCO subsidiary in the United States was Central Hudson Enterprises, a unit of Central Hudson Gas & Electric Company. These subsidiaries are unregulated and thus can earn profits in excess of the allowed rate of return of their parent utilities. In 1991, ten ESCOs responding to utility Requests for Proposals for DSM bidding programs were utility subsidiaries. It is generally recognized that these subsidiaries have certain competitive advantages including knowledge of the utility industry, access to customer information and ability to tap large amounts of capital at attractive rates. Because of this perceived advantage, regulators have been very concerned about "self-dealing:" a parent utility giving preferential treatment in a DSM bidding competition to its own subsidiary. For this reason, there has yet to be a single case of a subsidiary bidding to its parent, either because of regulatory prohibition or an unwritten understanding between regulators and utilities.

Pros and Cons of Demand-Side Bidding

With four years of experience with full-scale implementation of DSM bidding programs in one form or another, many utility industry observers have had the chance to form opinions about the viability of this particular approach. There has been great promise that DSM bidding could "level the playing field" between supply-side and DSM resources and provide utilities with an alternative mechanism to develop their DSM resources through the use of ESCOs. Nonetheless, there has been a rocky history of implementation with many suboptimal results. As a result, utilities, regulators, and ESCOs can all cite various pros and cons of DSM bidding [Wolcott, 1990 (1)].

Why Utilities and Regulators *Like* DSM Bidding

The performance contracting approach offered by ESCOs attempts to overcome a number of the perceived market and institutional barriers to direct customer investment in improved end-use efficiency options. ESCOs are a source of one-stop shopping for a range of capabilities that utility customers often don't have. For example, ESCOs provide engineering and managerial expertise which help customers assess and implement an optimal set of energy efficiency improvements. ESCOs also assume significant technical, financial, and operational risk in many projects because they typically receive a fee that is proportioned to the savings achieved. If there are no savings, there is no payment, and most ESCOs are willing to guarantee certain levels of savings. Finally, ESCOs arrange the project financing. Through DSM bidding, these capabilities (engineering/managerial expertise, risk management, financing) can be delivered to utility customers.

In several jurisdictions, regulators have been attracted to this service delivery model for various reasons (Goldman and Wolcott, 1990). First, some regulators view the relative successes of PURPA, which produced a viable independent power industry representing a competitive alternative to utility construction, as a model which can be applied to the development of DSM resources. These regulators would like to see the utilities face similar competition in the supply of energy efficiency services.

Second, in some states, regulators have been frustrated at utility reluctance to aggressively pursue DSM, particularly end-use efficiency improvements. Some regulators have apparently concluded that utilities are unable or unwilling to deliver DSM services. Under traditional regulation, utilities in many states still have little direct motivation to help customers use electricity more efficiently, although that attitude could well change to the extent that regulatory incentives are implemented. Moreover, some utilities have not developed the technical expertise to recommend appropriate efficiency improvements to their customers or do not have the administrative capability to deliver DSM services. A number of utilities are increasingly building these capabilities, in-house, through contractors, or through unregulated ESCO subsidiaries.

Third, in terms of DSM resource development, the long-term contractual relationships with customers and the explicit obligation to monitor retrofit performance are other attractive features of performance contracting. For example, an ESCO participating in a DSM bidding program may negotiate shared savings transactions with the utility's customers. Because the ESCO is often providing financing, it has a strong motivation to operate and maintain the equipment at a high level of efficiency in order to wring the cash flow out of the project required to retire the debt service, satisfy equity requirements, and make a profit.

This long-term motivation contrasts with many utility DSM programs in which customers simply receive a grant (as in a rebate) to install efficiency equipment, and are then left to their own resources—resources which are often lacking, and poorly motivated, to operate and maintain the equipment. Regulators hope that reliance on ESCOs will ensure the reliability and persistence of the energy savings, which some consider the potential Achilles heel of DSM [Wolcott, 1990 (2)].

Why Utilities and Regulators *Don't Like* DSM Bidding

The primary negative impression that utilities and regulators have of DSM bidding flows from the way ESCOs typically do business [ACEEE, 1990 (1)]. First, because of their interest in maximizing profits, it is generally perceived that ESCOs favor installation of short payback measures and seek to minimize their transaction costs. The predominance of efficient lighting retrofits in pilot bidding programs is often cited as an example of ESCO tendencies toward "cream-skimming," and their insufficient attention to implementing a comprehensive package of efficiency options in the customer's long-term interest. ESCOs have also reduced their marketing costs by concentrating on only the largest commercial and industrial customers, leaving the smaller ones unserved.

However, these problems can be minimized through effective program design, based on remedies developed by several utilities in pilot bidding programs. For example, Northeast Utilities' Contractor/Arranger program is designed to provide an incentive for only those measures with greater than a 3-year payback, relying on the customer and ESCO to arrange financing for the quick payback items. Some regulators have considered providing "mining rights," such that the ESCO can continue to exploit efficiency opportunities in its customers' facilities and receive the incentive payment from the utility, even after the original bid amount has been satisfied. The ESCO is free to operate under this arrangement until the next bid cycle is initiated.

Second, critics of DSM bidding argue that reliance upon third party firms makes the entire process too complex and burdensome for the utility and its customers. The ESCO must negotiate relatively lengthy and complicated performance contracts with utility customers, many of whom are inexperienced in these types of transactions, which require a difficult mix of technical, financial, and legal expertise to successfully negotiate. This problem often translates into extensive time required to move the projects to full implementation. ESCOs are keenly aware of this problem as the release of their security deposit is almost always a function of delivering their bid amount within a specified period of time. A utility could consider providing technical assistance to its customers in negotiating performance contracts. At a minimum, utilities could avail themselves of technical assistance services that many state energy offices provide to selected constituencies; e.g., public sector agencies, to procure performance contracts (TDC, 1989).

Third, the ESCO industry was plagued for years by an unfavorable image, which has created additional skepticism among some utility staff that are responsible for developing DSM bidding programs. The image problem goes back to the time when performance contracts were heavily tax-advantaged transactions, and there were a number of firms that were in business for the tax benefits and not the engineering services. In that era, ESCOs earned the reputation of "fly-by-nights" and "snake oil salesmen" with inadequate qualifications, experience, and resources. Then, with the demise of the federal tax benefits in 1986, there was the great shakeout which freed the industry of the snake oil but left some observers wondering if there was any industry left to speak of.

Fourth, a related concern often expressed by utilities is that the industry is under-developed and unable to provide enough ESCOs, with a track record, to ensure a competitive response to RFPs. This problem may be exacerbated by the increasing number of jurisdictions that are implementing DSM bidding programs. While this may be somewhat a regional phenomenon, utilities do have some remedies they can employ to mitigate the problem. For example, some utilities; e.g., Public Service of Indiana, have sought RFP responses from innovative sources such as customer load cooperatives that collaboratively bid to provide DSM services throughout the utility's service territory. Securing adequately qualified ESCOs may be more of a problem for DSM bidding programs that rely heavily on price, as compared to qualifications, in the selection criteria.

Fifth, some utilities are particularly concerned that ESCOs may get between them and their customers. When an ESCO solicits customers, does the customer view the ESCO as an independent vendor of energy efficiency services or an agent of the utility? If the ESCO is perceived as an agent, is the utility legally liable for its deficiencies? Even if the ESCO/customer contract clearly relieves the utility of any such legal liability, is there a public relations liability involved? Some utilities face this set of questions defensively and seek to isolate the ESCO/customer transaction from normal utility customer relations.

Perhaps more constructively, utilities could exploit the ESCO's capabilities and market them as a customer service. Toward this end, utilities can provide marketing assistance to ESCOs, can clearly communicate the program's benefits to customers, and can stand by the ESCO in its relationships with the utility's customers. However, the utility might be more willing to pursue this approach only if it has significant flexibility and control in selecting ESCOs with which it wants to develop a long-term partnership and business arrangement.

Why ESCOs *Like* DSM Bidding

The primary feature of DSM bidding that gets the attention of ESCOs is, of course, the financial incentive that the utility is willing to pay for the energy savings achieved. The ESCO can use the utility payments to leverage additional financing from debt or investment sources or, the subsidy might allow the ESCO to undertake longer payback measures. One can presume that the ESCO will pass through a portion of the benefit, especially if the customer is aware of the utility's financial incentive, allowing a project based on "shared subsidies" as well as shared savings. Such a project gives the customer a better deal than it would otherwise have received and, perhaps for that reason, makes deals happen that might not have been possible.

ESCOs cite the prospect of reduced marketing costs as the second most attractive feature of some utility DSM bidding programs. ESCOs recognize that DSM bidding programs can provide an important means to control these costs if the utility is willing to work aggressively with the ESCO to market the program. The assistance that ESCOs find most useful is the identification of utility customers that represent good prospects for performance contracting. Provision of a list of self-selected customers with the right person to contact and perhaps some customer load data will make the ESCO's marketing job easier. Working through the utility's program will also give ESCOs credibility they might not otherwise have.

Finally, an increasing number of ESCOs recognize that implementation of DSM bidding is opening up markets geographically to performance contracting that simply did not previously exist; e.g.,

Indiana. ESCOs are opening up new branch offices in the service territories of utilities implementing DSM bidding programs because they can now justify those expenses given the opportunities presented.

Why ESCOs *Don't Like* DSM Bidding

There is sufficient experience with pilot and full-scale DSM bidding programs to assess ESCO reaction to particular design features. While the ESCO industry is not monolithic and speaks in many voices, clear preferences have been presented by some firms for certain program design features (Bullock, 1990; Esteves, 1989; and King, 1990). However, regulators and utilities will obviously have to weigh ESCO preferences in the context of an assessment of ultimate benefits and costs to ratepayers and utility shareholders. Some specific ESCO preferences may have negative implications for the utility and/or its customers.

While most ESCOs acknowledge the desirability of security requirements to protect the utility from non-performance, some ESCOs believe that performance guarantees have been onerous in some programs. For example, the $15/kW requirement for New York's programs are generally acceptable to ESCOs, which contrasts with the opposition to requirements that were ten to twenty times that amount in the early New England Electric System Performance Contracting program. Programs that provide one-time, up-front payments are preferred by some ESCOs looking to minimize transaction costs and maximize the net present value of the utility subsidy. However, other ESCOs prefer installment payments over time especially if it's because the program is paying for kWh savings.

Some ESCOs prefer limited, if not exclusive, franchises, although there are significant policy implications to any approach that limits competition. Some ESCOs don't like intrusive impact evaluation requirements; e.g., submetering, no provision for sampling populations, especially if the ESCO has to exclusively bear the cost of complying with such data acquisition.

ESCOs generally do not like DSM bidding programs that place limits on eligible technologies or customer classes or in any way constrain the ESCO/customer business relationship.

ESCOs are exceedingly wary of the utility that does not appear to be fully committed to the successful implementation of its DSM bidding program. In deciding whether to respond to utility demand-side solicitations, ESCOs will consider whether: 1) the utility was ordered to undertake the program by its regulatory authority (bad) or the program is an integral component of its strategic DSM delivery plan (good); 2) the implicit time commitment made by the utility in designing the program; e.g., one year vs. three to five years; 3) administrative requirements that are perceived to be excessive; and 4) very low ceiling prices.

Regulatory authorities can provide important remedies to some of these problems. For example, utilities could be allowed to recover their DSM bidding program costs, including revenue losses where applicable, through mechanisms independent of the program; e.g., through regulatory incentives, rather than through reductions in ceiling prices.

Finally, some ESCOs are concerned that DSM bidding programs force them to do business unnaturally. For example, many bidding programs require ESCOs to specify the magnitude of energy savings at a competitively bid cost, for which it must make a security deposit, all without benefit of analyzing the customers' facilities. Normally, an ESCO would conduct an energy audit and then a detailed engineering analysis of a facility in order to understand and knowledgeably exploit the energy efficiency opportunities. However, DSM bidding raises the scale of operation from individual facility to service territory. It is extremely difficult to estimate technical or market potential accurately in a utility's service territory, unless the ESCO incurs substantial up-front marketing costs and has access to information that only the utility may possess. These risks can be reduced somewhat if the utility supplies billing data and existing audit/survey data which can be used by ESCOs in bid preparation.

Some ESCOs also have problems with the predominant electric efficiency bias of current bidding programs which tend to steer their attention away from other fuel efficiency opportunities; e.g., natural gas, that might be available.

Key Issues ESCOs Face in DSM Bidding

Demand-side bidding programs have not provided ESCOs with the opportunity for systematic entry into utility DSM markets that many had hoped for (Moss, 1991). ESCOs have faced a number of unforeseen marketing risks in the way bidding programs were designed and the decision process that utilities went through to select vendors.

Marketing Risks

A number of sources provide evidence as to how DSM bidding programs have been implemented in practice. These sources include the results of process evaluations, technical conference papers, and regulatory filings. The perspective of the ESCO industry is one of frustration. Many firms have submitted DSM bids only to be turned down and denied access to the utility DSM market. This situation has created substantial marketing risks for ESCOs that often invest $50,000 to $100,000 per proposal and look to participation in utility programs as an important way to establish themselves in new geographic regions. ESCOs have been rejected because of certain DSM bidding program design features and perceived competition between ESCO efforts and a utility's own DSM programs.

The most notable example of program design failure comes from Orange & Rockland's program in New York (ERCE, 1991). Through a number of influences, the program emphasized summer peak demand-reducing technologies, set an artificially low ceiling price for bids (accounting for the utility's revenue losses), and imposed a limited franchise such that only three ESCOs were selected for specific geographic regions within the utility's service territory. As a result a very attractive, cost-effective ESCO proposal was rejected. To make matters worse, when another ESCO, which had been selected subsequently withdrew because of failed contract negotiations, there was no mechanism to go back and accept the previously rejected proposal. This outcome in the New York portion of ORU's service territory was in stark contrast to the results in its New Jersey portion which did not operate under the restrictive program design features (Peters et al., 1991). In New Jersey, all ESCO bids were accepted.

ESCOs have also faced marketing risks as a result of the decision process employed by utilities to select winners. Most programs use a numerical system to score and rank proposals, often requiring that ESCOs self-score their own proposals based on unambiguous criteria. These scores are the basis for the "first stage" selection by which most ESCOs make it into the initial award group. However, many utilities then employ a "second stage" selection process based on qualitative and often unpublished criteria. There were two cases in 1991 in which only one or two ESCOs were finally selected from a dozen that had made it into the initial award group. In the case of the New York State Electric and Gas Company's program, the utility said that it had revised its demand forecast down substantially from one year to the next so that it no longer needed the DSM "capacity" despite having received a number of cost-effective DSM bids.

In the case of Niagara Mohawk Power Corporation (NiMo), only one ESCO was selected from ten that were eligible to serve all of NiMo's commercial and industrial customers. The decision process employed is noteworthy because it is well-documented by an independent evaluation and it was based on perceived competition between the ESCO bids and the utility's own DSM programs [LBL, 1991 (2)]. In sum, NiMo simply compared actual ESCO bids (in $/kWh) and rejected all that were greater than the cost of its own rebate programs. NiMo did this while imposing substantial performance and evaluation costs on the ESCOs that it did not bear for its own programs.

For example, ESCOs are required to guarantee the amount of energy savings that they bid for fifteen years. Such a contractual requirement forces the ESCO to either implement operations and maintenance procedures for that period of time or require the customer to do so. In either case, real costs are incurred that NiMo does not face because it imposes no similar 15-year performance requirement on measures installed through its own rebate programs. Similarly, NiMo imposed rigorous impact evaluation requirements that require the ESCO to meter a large number of end-uses to prove that the amount of savings bid are actually achieved. NiMo requires that the ESCO bear the full cost of the metering which is well in excess of the impact evaluation methods that it employs for its own DSM programs.

It is ironic that these features (long term savings guaranteed and rigorously monitored) are often used to highlight the unique benefits of DSM bidding programs in providing persistent and reliable energy savings that might not be assured through a utility's own programs. This benefit goes to the heart of integrated resource planning to the extent that DSM is viewed as a legitimate electric resource that utilities and regulators can count on over the long-term. In NiMo's program, this benefit was not only ignored, it was penalized.

The Future of ESCOs and DSM

These two cases do not appear to be unique. The contemporary model of DSM bidding apparently has substantial limitations from the ESCO point of view, and may even be called a failure. The basic premise of the contemporary model of bidding is that DSM resources are acquired in an integrated competition with supply-side resources in a selection process based largely on the bid price. Clearly, within the context of this model, utilities are limiting ESCO participation in their DSM programs.

There is also evidence that it is the model, not the ESCO, which is the problem. Some utilities recognize the substantial value added that ESCOs can provide, yet believe that the process of DSM bidding is deficient in giving them the control that they feel they need to hire and deploy ESCOs to their customers. These utilities are engaged in defining new programs that go beyond bidding to bring ESCOs into the mainstream as an integral component of the utilities' DSM program.

This new generation of ESCO program is modeled on a relatively old one, the Contractor/Arranger program of Northeast Utilities (NU), in which ESCOs are hired competitively based on their qualifications and a price bid to provide engineering and construction management services on a time-and-materials basis. NU provides a payment equal to the ESCO's fee plus the full cost of DSM measures with greater than a 3-year payback but less than an amount that just passes NU's Total Resource Cost Test (around a 5-year payback). The new model ESCO program (being tested by the Potomac Electric Power Company) takes NU's program and adds an incentive for the ESCO to enter into a long-term performance contracting relationship with the utility's customers. This approach may be the combination that gives the utility the control it needs while taking advantage of the substantial service and value-added that ESCOs provide in acquiring DSM resources.

CHAPTER 17

AN ENERGY SERVICE COMPANY'S PERSPECTIVE ON DEMAND-SIDE MANAGEMENT

Cary G. Bullock

In 1985, Massachusetts Electric Company held a bid for demand reduction to be supplied by energy service companies (ESCOs). The effort was of interest because it was one of the first demand-side management (DSM) bidding programs held in the United States. Since then, several DSM auctions have been held. Many, particularly those in the environmental community, expected DSM bidding to be a key component in integrated resource planning. Many observed similarities with supply side bidding, and expected DSM bidding to grow accordingly.

Today, more than $2 billion annually is being spent by utilities on DSM programs. Less than $100 million is spent with ESCOs. Does this mean that ESCOs provide a type of DSM which is not useful to utilities or that ESCOs and utilities have difficulty working together? Although the experience has been mixed to date, there are genuine benefits to utilities, their shareholders and ratepayers in working collaboratively with ESCOs in DSM implementation. This chapter explores some of the reasons for the current situation and suggests some alternatives, which could benefit all of the players.

The Traditional Business of The ESCOs

The National Association of Energy Service Companies (NAESCO) defines an ESCO as a company that provides the financial investment for an energy cost reduction project, and receives a share of the benefit created by that project, as compensation for the investment made. The contract vehicle used by ESCO's goes by several names, such as shared savings, energy purchase contract or performance contract. For ease of reference, I shall refer to this mode of business as performance contracting.

Historically, this method of doing business has had several benefits for the project host. First, it has obviated the need for the host to provide investment capital. Second, it has provided the host customer a qualified contractor, who understood the technologies and how to get them to perform. Finally, it permitted the host facility to realize savings, which may not otherwise have been realized, because the investment hurdle rate of the ESCO has often been lower than that of the host.

Historically, there have been three problematic aspects of performance contracting. First, measurement of energy savings has often been difficult. Second, the contracts required are complex and are new to prospective customers. Finally, hosts must be adequately credit worthy so they can be expected to remain in business for the term of the contract. All these factors increase the transaction cost to the prospective customer.

Measurement difficulties have been mitigated by ESCOs over the years by moving from a total energy baseline measurement approach to measurement only of specific equipment that was modified. Contractual complexity has been mitigated by gravitating toward standard contract vehicles which resemble leases, but have been modified to recognize the contingent nature of the payments. Credit issues were largely ignored by ESCOs in the early days of the industry. Today, smart ESCOs evaluate credit in a manner similar to a bank contemplating a loan to a customer.

ESCOs have participated in utility DSM programs in two modes. First, utilities have sometimes awarded contracts which were structured as traditional power purchase contracts. In these instances, the utility contribution has been coupled with the savings stream to justify the upfront capital investment and thereby creates a contribution to project capital. Second, utilities most often have provided rebates to customers, which do not fully pay for the project. The rebate is used, along with capitalization of the potential savings stream, to justify the upfront capital investment.

ESCOs and Utilities See DSM Differently

ESCOs and utilities have a different perspective on DSM. To the ESCO, DSM is an extension of a prior business; that of selling savings produced on a unit basis as, and if, savings are produced. Their approach to DSM is analogous to building a "conservation power plant." To the ESCO, DSM is a long term relationship to be maintained.

To the utility, DSM implementation is simply equipment deployment. For most utilities today, the assumption is that good equipment, properly installed, can be counted upon to produce savings over the life of the equipment. Measurement, therefore, is less a matter of looking at any one facility; rather, it is a matter of evaluating the relationship of the total installation on a statistical basis long after the fact. To the utility, the relationship with the host DSM project is generally short term.

For the utility, DSM is potentially many other things. For example, since many "outsiders" often see DSM as an entitlement program, deployment of DSM funds may have political value to the utility. DSM is also a powerful marketing tool, for it permits a utility to distribute rewards to favored customers. Because regulators have begun to place a high value on DSM, it is also important during rate cases.

Utilities must deal directly with several problems which have political overtones. Many are concerned that utilities will pay for conservation, which would have occurred anyway (the "free riders" problem). Others are concerned that one rate class not subsidize another rate class, or that one group of customers benefit to the detriment of other customers (the "equity" problem). Still others are concerned that DSM within any given facility be comprehensive. All these issues complicate seeking least cost DSM, because such constraints act to increase costs.

ESCOs, in general, do not see the broader side of DSM as the utilities see it. They tend to finance their projects from the expected cash stream from host customers and/or utilities. Utilities generally finance these core DSM program expenditures by adding program costs to the rate base as costs are incurred. Over and above total cost recovery, a utility may also receive an incentive, which is a percentage of the future benefits expected to be realized by ratepayers. When this is the case, the utility is essentially a large ESCO, having a shared savings relationship with the ratepayer. The ratepayers, who accept a higher rate for DSM performed by the utility, receive the benefit of not having the utility incur the cost of acquiring additional power.

For the purpose of the discussion which follows, I refer to the traditional ESCO approach as the "performance paradigm" and the traditional utility approach as the "core program" paradigm.

SOME DIFFERENCES

The primary difference between the performance paradigm and the core program paradigm lies in who bears the risk for nonperformance over time. What happens if savings do not persist? In the performance paradigm, this risk is borne by a combination of the host customer and the ESCO. In the core program paradigm, this risk has historically been borne by the ratepayer.[1]

A second difference is that the core program DSM tends to be passive when problems occur. If, for example, savings do not materialize, there is no mechanism to tell exactly where the problem lies. This result is due to the fact that this type of DSM is evaluated on a statistical basis long after the fact. One may be able to characterize shortfalls *ex post facto*, but it is difficult to know which facilities have problems and to mobilize efforts to resolve those problems. Moreover, once the problems are known, the opportunity to correct problems may have less immediacy with hosts.

A third difference is the way programs are financed. Performance paradigm DSM covers all contingencies sufficiently well; so it can be financed on a non-recourse basis. This simply means that projects may justify their own financing without outside help. This, in turn, also means that the financing may be leveraged with long term debt to reduce

---

[1]Boston Edison has sought to create a hybrid for the performance paradigm and the core program paradigm in its ESP DSM program. In this program, the customer is required to take the performance risk for two years. In return, the utility agrees to pay the project down to a one year payback.

the finance costs. In core program paradigm financing, this is not now the case. Currently, core program is essentially a pure equity financing, in that it is paid directly by rate payers in their rates; i.e., it is not rate-based. Fortunately for utilities, when ratepayers provide this type of equity today, they do not ask for market returns, so it is inherently less expensive than market capital.

Some Misconceptions

Several misconceptions arise relating to DSM as delivered by these two different mechanisms. First, many believe that only the utility can deliver DSM via the core program paradigm. This is false. In principle, any entity, with access to the ratepayer finance mechanism as well as utility customer databases relating to usage, can perform the same function as the utility. Indeed, it is not clear that the utility is either the best, or least cost mechanism, for delivering DSM, regardless of the delivery paradigm.

Second, many feel that DSM delivered via the performance paradigm and the core program paradigm is identical. In fact, this is also false. Under the performance paradigm, the customer and the ESCO undertake the obligation to ensure that savings persist over time. Indeed under many contracts, there are significant penalties in the event savings fail to persist. DSM developed under the core program paradigm generally requires the ratepayer to take the risk of long term persistence. In essence, savings are what they are. This factor is one of the significant weaknesses to the core program paradigm. If savings fail to persist over a long period of time in any significant manner, the ratepayer will generally have overpaid for the savings. Not only will they have paid for the costs of the DSM program, but they will also have had to pay the costs incurred by the utility in buying power, which would not otherwise have been required. DSM, under the performance paradigm is a firm resource, whereas DSM under the core program paradigm generally is not.

A third belief held by some is that DSM developed under the core program paradigm is inherently less expensive than that developed under a performance paradigm. The answer is simply, "It all depends." One of the cost advantages core program DSM has over performance

DSM is that the risk of non-performance is not priced. If reality never calls the risk; i.e., if in the specific program, the savings persist over the life of the equipment, then their costs are truly zero. It is much the same as saying it is less expensive for you not to have life insurance over a specific period, if you don't die. On the other hand, if savings from DSM programs evaporate over time, the core program paradigm gives no recourse for mitigating damages to the ratepayer. DSM developed under the performance paradigm provides protection against this eventuality. If savings degrade over the life of the equipment, core program DSM can be much more expensive than performance DSM.

A fourth misconception is that it is too costly to measure savings. While it is true that not all savings can be measured, much more can be measured cost effectively today than five years ago. If measurement is required, costs to gauge savings will continue to fall precipitously in the future, as they have in other areas.[2]

A final misconception is that there are not enough ESCOs to provide the DSM required by utilities in the time required. While this view has been successfully rebutted, it was nonetheless a strong argument when power shortages were immediate. Two comments are in order. First, the early 1990's recession in the United States bought considerable time for growing the industry. Second, if companies who take these risks are permitted to earn the economic returns, normally due to those who solve problems early and well, then business history would suggest that the business community can and will respond.

To date, the problem has been that the process offered by utilities to third parties is simply much more complex than it needs to be. Moreover, there is general dissatisfaction among purchasing utilities to pay prices associated with the risks undertaken as defined by bidding in the marketplace. Most wish to pay only for the equipment installed, rather than the value of the commodity (savings) produced by the equipment. So long as this is true it will be necessary to have ratepayers

---

[2]Measurement costs are largely the costs of installation and instrumentation. In the last twenty years, the costs of primary computer memory has fallen 50% per annum. The costs of computation power has fallen even more dramatically. This author is aware of the proprietary development efforts which would reduce the cost of instrument installation dramatically. Finally, some manufacturers are now installing savings monitoring devices directly within new equipment. All these factors suggest that direct measurement of savings will continue to fall.

not only pay for equipment, but provide much of the financing to utilities as well.

Flashpoints between Utilities and ESCOs

ESCOs have responded to the needs posed by utility DSM programs. This experience has been mixed on both sides.

First, as regards to DSM bidding, ESCOs, in general, have responded well. In most instances, at least three times as much was bid as was sought by the utility. In some recent cases, however, utilities have run lengthy and complex bidding programs, only to select very little of the offering. Several issues continue to surface. In the Niagara Mohawk DSM bid of 1990, a single ESCO was selected. The two reasons cited by the utility for refusing to buy any more from the offering were: (1) the prices offered by bidders in general were much higher than the utility wanted to pay; and (2) if the programs offered by bidders were accepted, it might compromise the utility's ability to complete its own core programs since some bidders sought a similar market.

From the ESCO perspective these were, and are, valid reasons. The objection from the industry, however, is that these issues could have been addressed in the RFP and were not. ESCOs spend considerable money preparing a bid; hence, if a utility is not serious about a solicitation, or if a utility doesn't know what it wants, it is wise to pass on the offering.

In a later auction in the same region held by New York State Electric and Gas (NYSEG), some ESCOs elected not to bid. First, NYSEG appeared to have designed core programs, which dealt with most sectors of their service territory. Second, some felt that the utility industry in New York was less than enthusiastic toward using third party ESCOs in a bidding context. Third, some ESCOs felt that the utility would use a different selection criteria than had been publicly stated. The results of the auction seemed to justify these concerns.[3]

The net effect of such experiences is that good bidders tend to abandon the market. This, in turn, hurts the ratepayer, because it

[3] At least two ESCOs intervened to have the Niagara Mohawk awards set aside because of unfair practices by the utility. The same thing happened with regard to the NYSEG bid.

eliminates options which should be considered.

An area of utility concern is that the utility cedes much of the control of the DSM program to the ESCO. Utilities are inclined to want the last word with regard to a customer project. For ESCOs this is problematic, since the ESCO's bank or investors feel they have earned that right by virtue of providing the investment capital for the project. This may bring them into conflict with the utility, because the utility may not be indifferent to the variety of options a given customer may have for DSM.

A second area of utility concern is that not all experiences with ESCOs have been positive. Not all ESCOs have met their contract obligations. The nature of performance contracting is that some will fail. The industry is young and is improving in both its capitalization and delivery capability each year. In many cases, ESCOs have taken the risks for which utilities later have earned profits. The industry has earned the right to be a player; and, as a player, it will improve if allowed to play.

A third area of concern is that utilities and outside groups are afraid that performance paradigm programs are simply too difficult to market and experience to date tends to confirm this. Rebate programs generally require no downstream customer obligation while performance paradigm programs do. Customers are more likely to accept rebates unless a significant premium is offered for a term commitment.

A fourth area of concern is that ESCO programs will compete for the same customers as core programs, and reduce the effect of those programs. It is easier to control the effect of core programs if one only uses ESCOs as contractors, or if one restricts the ESCO access to the market. This issue is complex and may need regulatory guidance for resolution. Hopefully, it will be decided on the basis of the best value for the ratepayer.

A final area of concern is competition for profit. Increasingly, utilities are functioning as large integrated ESCOs for their service areas, albeit in a monopoly posture. Today, regulatory bodies do not require utility shareholders to take the level of risk, which utilities accord ESCOs or ratepayers. If this continues, ESCOs will always have a disadvantaged position in the market. Utilities often see themselves competing for the same profit margin as the ESCOs. Hence, some believe they will have greater profits and more control if ESCOs are constrained in the market or eliminated entirely from it.

Do ESCOs Offer Value To Utility DSM Programs?

There are several areas where utilities and ESCO can work in a complementary way in DSM programs.

First, from the history of DSM it is clear that the utility shareholder, as evidenced by management recommendations to regulatory bodies, does not wish to take any significant performance risk on DSM programs. As of this writing, I know of no utility core program wherein the utility is asking its regulators for the right to invest shareholder funds and to have these paid back as the savings materialize. Most, in fact, wish to be paid for future benefits today. In contrast, ESCOs have developed a variety of financial sources, who seek to take this risk, whether with individual customers or with programs as a whole. Historically, utilities have done business as monopolies. When it comes to DSM investments, utilities would do well to subject themselves to the same due diligence that ESCOs subject their DSM investments to. Utilities should proactively seek out ways to get ESCO shareholders to take the performance risk in their programs. I believe this could be done, on a customer by customer or on a programmatic basis.

From an ESCO perspective, if utilities want to invest in DSM for the returns available, they should consider investing strictly as investors and not as active operating managers. Most ESCO investors do just that. They are seeking investments with people who know how to manage projects and who have built the infrastructure to manage the projects. Indeed, they wish to distance themselves from the day-to-day operational aspects of the business. Most ESCOs are not capitalized to own projects over the long term. Utilities, who wish to make this kind of investment, should consider investing in other people's DSM projects.

While ESCOs know much about specific technologies and specific customer niches, utilities are far better at providing broad marketing programs for their customers than are ESCOs. Utilities and ESCOs should work together in this area and ESCOs should respect the reasons utilities need to take the lead.

Certain types of risk and activities are inherently less expensive for the utility. For example, the credit risk is inherently less expensive to the utility (ratepayer and shareholder) than it is for the ESCO. The reason is simple: while a given customer may go bankrupt, the facility does not. Moreover, many facilities are not renovated when a new tenant moves in. Utilities also know more about the customer's payment patterns and credit capability than does the ESCO. Utilities are far more experienced than are ESCOs in billing and collecting from customers. Ironically, this is a service ESCOs might willingly pay for, and it is also a service which would offer the utility an opportunity to show the effect of the program to the specific customer.

Finally, ESCOs are better at staging comprehensive projects over time than utilities because of expertise they have developed. ESCOs are generally willing to spend the time and invest *(and risk)* engineering dollars to convince customers of comprehensive approaches. Utilities are not. Utilities should find ways to harness this expertise.

SECTION 3

THE INSTITUTIONAL SECTOR

The institutional sector is comprised of public and private nonprofit organizations, including public schools, hospitals, public care institutions, local governments, colleges and universities. Since public schools operate under rigid budgets and generally confront the most adverse conditions related to energy costs, they have been selected to explore how performance contracting works in the institutional sector. While the focus is on schools, the concerns and procedures addressed here generally apply to all public and private non-profit institutions.

No one knows better than the school administrator that institutions are facing a financial dilemma, made worse by vacillating energy costs. Schools operate on budgets adopted by their boards and/or governing bodies. The revenues the schools have available for operation are often set by state legislative appropriations, and they were allocated one or two years previous to any board action. Schools, therefore, have little financial flexibility to meet volatile energy prices. When prices rise unexpectedly, as they did in the fall of 1990, school districts generally have to cut educational programs or services to pay the utility bill. To make matters worse, the public schools are the *only* sector of the economy that cannot raise the price of its goods or services to pay the utility bill.

Historically, school administrators have paid the fuel bill by dipping into funds allocated to operate the facilities, particularly the maintenance budget. Ironically, poor maintenance reduces energy efficiency and the utility bills climb even higher. Furthermore, energy costs frequently claim the money that might be used to improve the building envelope or buy more efficient equipment. Poor maintenance is also a major contributor to indoor air problems. As a result, the public schools tend to get mired in a progressively less energy efficient operations and growing indoor air concerns.

The alternative, of course, is for the administration to take definite steps to conserve energy.

Several avenues are open to the schools to find technical and financial support to cut operating costs. The local utility may offer rebates or other incentives that help finance energy efficiency measures. They may also provide audits or technical consultation. Local firms, such as architectural and engineering firms, may provide expertise or support as a gesture of community support, or for a share of the savings.

Many districts have taken advantage of the U.S. Department of Energy's Institutional Conservation Program (ICP) energy grants made available through state offices as discussed in Chapter 20. These grants, however, require local matching funds; and, unfortunately, many school systems have been forced to turn away from this assistance because they lacked the matching funds. When Congress reauthorized the ICP grants program in 1990, it expressly provided that performance contracting could be used as a match by institutions.

Whether ICP grants are involved or not, performance contracts offer school districts a source of financing and expertise for a comprehensive energy management approach that frequently exceeds the support available from other sources.

Chapter 18 offers an excellent example of the considerations and procedures one system, Charles County Public Schools in Maryland, faced in electing to use performance contracting. The case study is taken from a paper Mr. Joseph Lavorgna presented to the 14th World Energy Engineering Congress, October 1991. It is reprinted here with the permission of the Association of Energy Engineers.

In Chapter 19, Mr. R. Scott Holland, Honeywell's National Sales Manager for Schools and Colleges, looks at performance contracting for school districts from the ESCO's perspective. The fact that the firm selected by the Charles County Public Schools as its ESCO happens to be the one Holland represents is coincidental and should not be construed to suggest anything more. Holland's observations do not pertain to Charles County, nor are his remarks offered in response to Lavorgna's observations.

Matching the intricacies of performance contracting to a federal grants program calls for some skillful maneuvering. The last chapter discusses the general procedures for making the ICP/performance contracting marriage work. Specific procedure may vary over time and from state to state, so public administrators and ESCOs should consider Chapter 20 a general guideline and check with the state energy office in which the planned project is located before taking action.

CHAPTER 18

AN ENERGY SERVICES CONTRACT WITH GUARANTEED SAVINGS THAT REALLY WORKS

Joseph J. Lavorgna, P.E.
James A. Smoyer, P.E.

BACKGROUND

- Board of Education of Charles County, Maryland
- Suburban/Rural Public School System
- 461 sq. mi.—30 miles south of Washington, DC
- Enrollment approximately 20,000
- 30+ Buildings (2.2 million sq. ft.)
- Oldest building 1966
- Budget $86 million
- Energy Budget $2.8 Million
- Electricity $2.3 M
- Fuel Oil $.5 M

Charles County Maryland, as indicated in the background box, is a suburb of Washington, D.C. and as with other jurisdictions surrounding the nation's capital, it is experiencing long term growth. Currently Charles County is one of the fastest-growing areas in the region and its school construction program is desperately trying to keep pace with the number of new students.

The Board of Education of Charles County serves approximately 20,000 children enrolled in the county's only public school system. Charles County's public school system is the second largest employer in the county with more than 2,000 full- and part-time employees. The Charles County public school system operates 16 elementary, six middle, and four high schools. Additionally, the public school system offers specialized services through a vocational-technical center, a special education center, a central office/warehouse building and several other small ancillary facilities. Approximately $2.8 million of the system's $86 million annual operating budget is earmarked for fuel oil and electricity (natural gas is available only in a small service area in the county).

Why We Chose to Purchase a Guaranteed Savings Energy Services Agreement?

In the late 70's, the school system took advantage of the "Schools and Hospitals" grants program [ICP] established by the Department of Energy. We conducted preliminary energy audits and energy audits on all eligible schools. We implemented maintenance and operations measures to the best of our ability. We also acquired some energy conservation measure (ECM) grants to add insulation and to install time clocks. These measures were valuable and helped reduce our energy consumption. What the program really did best was to raise our understanding of which types of ECM's worked well and which types didn't have a good return.

One of the best investments we made was the installation of a WattShaver energy management computer in a few of our buildings. Through the Schools and Hospitals Grants program, we were able to add schools to the system until we had a total of 15 schools controlled by the computer. Over an 11-year period, beginning in 1977, we were able to reduce the energy consumption (MBTU/sq.ft.) in school system facilities by approximately 30%. (See Figure 18-1.) This reduction was accomplished by a combination of strategies employed in both computerized and non-computerized buildings.

The question is then—if we were so successful in reducing consumption, why were we looking to a guaranteed savings contract? The main reasons were:

- Our buildings were growing older and the mechanical equipment was beginning to break down more frequently. (We had little in the way of a preventive maintenance program.)

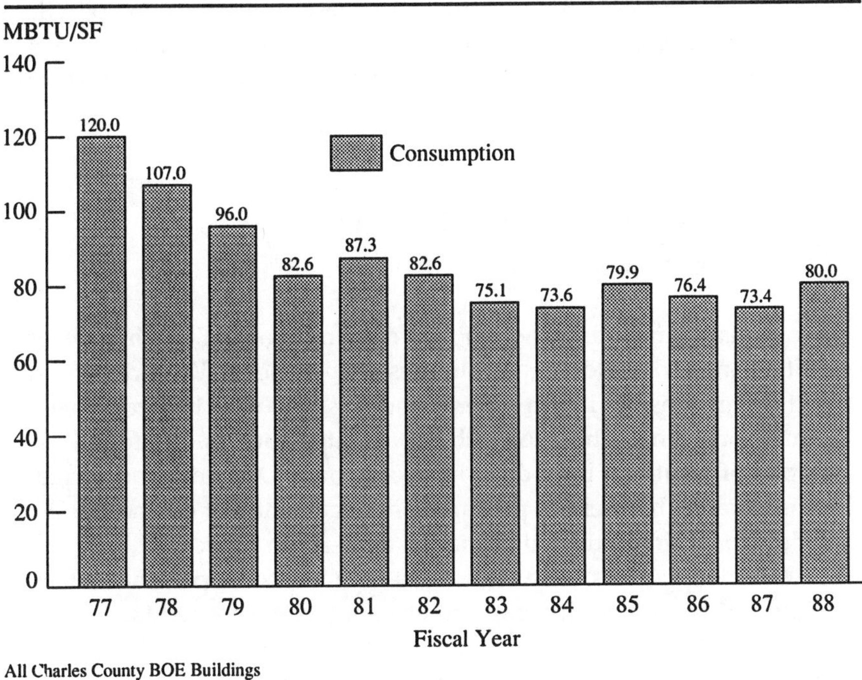

Figure 18-1. Annual Energy Consumption; 1977-1988

- We had great difficulty in securing funding for maintenance of equipment. (We never had a temperature control service contract on any of our buildings.)

- Rapid improvements in computer technology had made the Watt-Shaver computer system obsolete.

- Our staff was committed to improving our energy performance and, more importantly, concerned with the deteriorating comfort levels in our buildings. (Parental and staff complaints persisted.)

- We were dreamers about what could be done—we had a vision of where we wanted to be, and we were naive about the problems we would encounter.

- We thought we had nothing to lose. We had already reduced our consumption by 30%. Writing a Request for Proposal to upgrade our systems and pay for the work out of energy savings seemed like a sure bet. So, in the spring of 1987, we wrote a Request for Proposal (RFP).

What Was in The Request For Proposal (RFP)?

What to include in the RFP was our first problem. We had never written one and couldn't go to our files to dust one off. We did the next best thing; consulted with other school systems who had completed similar programs. This met with very limited success. We found examples of what had been done in Georgetown, Delaware and Media, Pennsylvania, but neither procurement document met our needs. We ended up writing it the old fashioned way—from scratch.

The basic goals were clear to us—we wanted a program which addressed people, equipment, and technology. We needed to:

1. Upgrade, repair, and/or replace all temperature controls.
2. Repair all HVAC equipment.
3. Replace our obsolete computerized energy management systems (EMS) with a state of the art system in all of our buildings.
4. Train our personnel.
5. Do all of this with little or no risk to the school system.

We had no idea of what to expect in terms of what would be proposed or who would be submitting proposals. What we did know was that we wanted to provide the greatest potential for savings within our operations. So we decided to allow each respondent to the RFP to create the best package of improvements they could offer in terms of proposed work versus savings generated. We suspected that proposals of this nature would vary greatly and would be very difficult to evaluate. It was decided to establish the evaluation criteria and publish them as part of the RFP. This was a difficult step but an important one. We opted for the following criteria and weighting:

- corporate fiscal and employee capacity to meet proposal requirements—20%
- experience, demonstrated ability, and reputation in similar projects—20%
- guaranteed energy savings and maintenance program content—30%
- estimate of savings—30%

All vendors were given six weeks to survey the buildings and to obtain firsthand knowledge of the condition of our equipment. As it turned out, this amount of time was barely adequate to perform surveys in 27 buildings and to develop a comprehensive energy savings proposal. However, it did serve to weed out those who did not have the resources to complete the surveys.

At this point, we were a little nervous about how we were going to manage this whole process given the fact that this was something we had never done before, and that we had other responsibilities that kept us busy the other 110% of the time.

The school system hired a consultant to lead us through the "mine field" we suspected we were walking into. (Besides, conventional wisdom holds that if something goes wrong, you can always blame it on the consultant.) The idea to hire a consultant was not a move which was received with open arms at all fronts. The recommendation persevered, however, and we engaged the services of Yewell/Energy of Lutherville, Maryland after the RFP was in the hands of prospective respondents.

At times during the survey period of the RFP, it was difficult to appreciate the value of having a consultant on board. After all, his work was not clearly defined (unlike an architect designing a building). Most of the decisions had to be made by the staff, but after a while, it became obvious to us that he was the glue that held the project together. He was very active setting up regular meetings, keeping minutes of those meetings, gently reminding us that we had deadlines to meet, and we drew on his expertise to evaluate the responses. His services were invaluable and the project might not have survived the many pitfalls we encountered had that objective outside voice not been there. **A key to the success of any agreement of this type—get some good outside consulting help.**

However, the best was yet to come. In addition to John Yewell, who handled the administrative and coordinating chores, John brought in James L. "Jim" Coggins, P.E., of Energy Applications, Inc. of Columbia MD, and Anthony M. "Tony" Carey, Esquire, of Venable, Baetjer, and Howard, of Baltimore, MD. Jim Coggins has a Ph.D. in mechanical engineering from MIT and Tony Carey is an attorney who specializes in performance contracts. Both of these individuals brought highly specialized experience to the team.

After we received proposals and interviewed the three firms who responded, a screening committee made up of staff members from facilities, maintenance, operations, engineering, finance, purchasing, and the consultant. We selected Honeywell because of the quality of their proposal. We were not leaning toward Honeywell before the screening. In fact, the facilities staff would have preferred a different vendor because of observations made during site visits to other school systems using a competitor's system. In retrospect, the value of the composition of the screening committee was to remove favoritism and prejudgement from the selection process.

What followed was six months of negotiations with the successful vendor before we were able to come up with an acceptable contract. This was both a stressful time and a productive time. It was during this period that the understandings and philosophies which were to form the foundation of our partnership were hammered out. These included:

- A respect for the needs of each other—we both had to be winners if this long term contract was to survive.

- There had to be full commitment from both sides to devote the resources necessary to keep this project viable.

- We developed a comprehensive contract that was understood by both sides up front.

- Anticipating that our contract was not a perfect agreement, provisions would be needed for growth and adjustment.

It was also during this period that the strategies for reducing our energy consumption, while maintaining comfort levels, were established. We agreed on: the method of calculating energy saved, implementation schedules for all phases of work, established priorities, types of lighting retrofits, the addition of security points to computerized Energy Management System (EMS) points being monitored. Wishes and promises met head-on with healthy doses of reality and skepticism. Each member of the team maintained their company loyalty, but also developed a commitment and loyalty to the project. Our consultants earned their fees during this phase and deserve much of the credit for the shape of our final agreement.

Indoor Air Quality (IAQ) Concerns

Enter indoor air quality (IAQ) concerns. While all of this was going on, the school system was jolted by two incidents which verified how much we needed this work to be done. First, students walked out of our largest high school complaining of allergic-type reactions: headaches, burning eyes and sore throat. Mold also was discovered in the library. Second, a severe roof leak during reroofing operations flooded 2/3 of a fully carpeted open space elementary school. Similar symptoms developed and parents were threatening all types of actions.

What we uncovered, while trying to determine the cause of the allergic symptoms, was the true condition of our mechanical equipment. Much of it suffered from benign neglect due to budget constraints—outside air dampers were inoperative, coils and filters were clogged, and compressors and fans were near the limit of their useful life. We quickly discovered the two schools were just the tip of the iceberg. We needed help, we needed it fast and we were under the scrutiny of the entire community to correct the problems, so the school system could get back to its main mission, educating students, not conducting press conferences. We were fortunate to have the solid support of our elected board and county commissioners in terms of funding to do something about it.

Honeywell agreed to commit many manhours to correct the deficiencies in dampers and controls at the two schools, even though we did not yet have a signed contract. In return, we agreed to pay the overtime differential to get the work done as quickly as possible. That type of commitment and compromise helped cement our relationship and typified the give and take in the negotiations phase of our contract.

After an intensive clean-up effort and a collecting of many air samples, ventilation was restored to normal levels. The allergic symptoms subsided, hysteria was eventually replaced by skepticism, and we learned a painful lesson—quick and open responses are vital in handling IAQ complaints. We also learned that in addition to all of our temperature control problems, we needed to budget some money for air balancing and testing. We had shot ourselves in the foot by building partitions and enclosing open space classrooms without adequate consideration for the heating, ventilating, and air conditioning (HVAC) systems.

Contract Signing

Finally, on February 9, 1988 the Board of Education signed a 6-year energy services agreement guaranteeing savings of $468,232 per year, financed by a third party municipal lease with Citicorp. The agreement included:

- $1,522,100 in repairs and modernization of controls
- Lighting retrofits—8,000 current limiters—replace 750 exit lights, convert 500 incandescent to fluorescent lights
- Centrally monitored computerized energy management system with 1,166 controlled points
- 160 hours per year of staff training
- Maintenance contracts for all temperature controls and the computerized EMS system
- $45,900 to add security reporting capability of 290 security points
- $34,500 for maintenance software capability

- A telephone service link to monitor our EMS alarms 24 hours/day (including refrigerators and freezers)
- Our personnel retained control of operations of EMS.

Contract Implementation

The signing of the agreement was only the beginning of the *real* job of creating a viable working system. The implementation phase took over 12 months to complete, and the start of the guarantee period was deferred until July 1, 1989. This allowed a shakedown period for the system and coincided with the beginning of the fiscal year.

One of the toughest problems we had to overcome was establishing a mutually agreeable base year of energy consumption for calculating energy savings. This, we assure you, is not as simple as you would think. For one thing, we had changed programs in many of our schools subsequent to publishing our RFP. We added six day care centers, many community recreation programs and some summer school programs. Second, our utility bills for the years prior to the RFP reflected poor comfort levels because (unknown to everyone) a large portion of our mechanical equipment was inoperative. This gave a false picture of our consumption patterns. Finally, we were collectively incapable of sorting out all the additional costs for operating the new programs and the lost savings caused by repairing all of our inoperable mechanical equipment. We raised the white flag on trying to reconcile our prior consumption patterns against current usage. We also knew the importance of restoring comfort levels and maintaining ventilation standards.

This was a major dilemma that had the potential to unravel our agreement. Both we and Honeywell had too much at stake to let this problem stand in our way. To resolve this issue, we agreed to establish a new base year by auditing our utility bills during the period after the equipment was repaired and before the computerized energy management system (EMS) was fully operational. The amount of the guarantee would not change, and the base energy consumption data would accurately reflect our current operation. It took almost a full year to install the EMS, test each location, and enable EMS software strategies system-wide after repairs were made to all the mechanical equipment.

The 17 months between the signing of the agreement and the beginning of the guarantee period were filled with activity. We had monthly progress meetings which were conducted much in the same way a construction progress meeting are run. Early in this phase, a schedule was established for each component of the work—repairs to equipment, control upgrades, lighting retrofit, EMS installation, and training. Our consultants played a major role in this phase by reviewing shop drawings and repair proposals, inspecting work in progress, and helping to establish control strategies for control retrofits.

A tremendous amount of coordination was required during this time. Each school had as many as four separate crews working on different parts of the project. Each crew had to be scheduled and the school principal advised so as not to disrupt the instructional program while work was performed; e.g., most of the lighting work was done in the evening, on weekends, or in the summer.

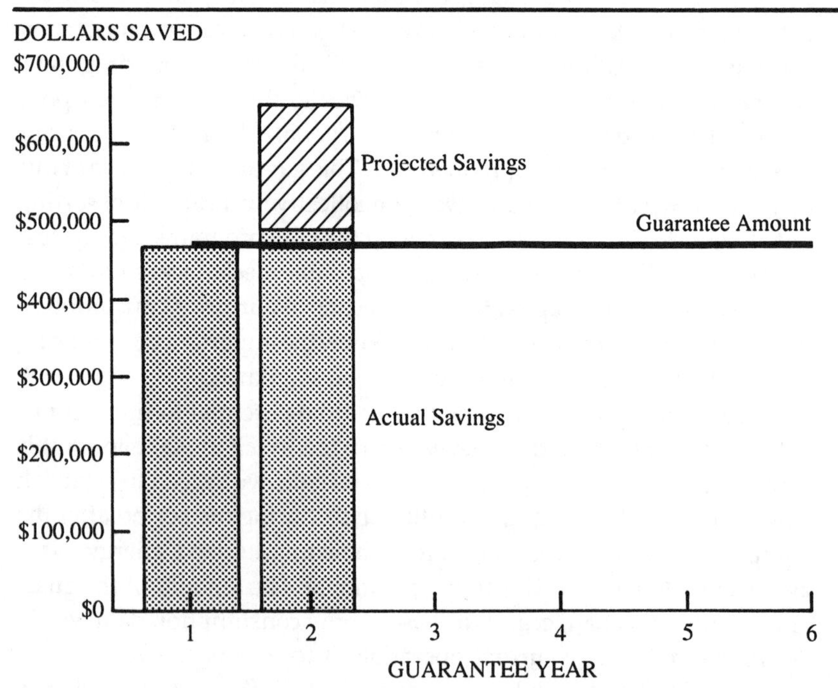

Figure 18-2. Guaranteed Savings Performance

To facilitate communication, we set up an office space for the Honeywell project manager who was on site four days each week. Much effort also went into informing the individual school staffs as to what the system could do and what it couldn't do. This was done by attending staff meetings, making presentations, and answering questions in each school that requested a briefing.

After all the work was completed and accepted, we signed a release authorizing payment to Honeywell of $1,552,100 on March 1, 1989.

Results of The First Year of Operation

After our first full year in the guarantee period, we were pleased to learn that the program had saved $471,229 approximately $3,000 more than the $468,232 guarantee. (See Figure 18-2.) This was a great relief. We knew if our vendor lost big money in the first year, and if that trend was projected to continue, our agreement would be in trouble. Fortunately for both of us, this did not happen.

In fact, through the first three quarters of the second year of the agreement, at the time of writing this paper [June, 1991], we will have saved $493,963. This savings has been used to help cover a budget shortfall caused by the current recession. We are projecting a total savings of $190,000 over and above the guarantee for the 1991 fiscal year.

Where Are We Now and Where Are We Headed?

Many things have changed since we signed our agreement three years ago. All but one of the people in the vendor's branch office, who were part of negotiating our agreement, have been promoted or transferred, but there has been no change in commitment to us as a customer. Computer technology has taken another quantum leap. We are upgrading our EMS computer as part of adding a new high school building as it is being constructed. Three other elementary schools which have been built or renovated since we initiated the RFP have not been added to the system due to lack of funds. These may be added by financing them with a similar lease purchase agreement. However, since they are new and energy efficient, it may be difficult to do.

We now receive very few complaints about comfort in our schools. Also, our indoor air quality complaints have likewise been drastically reduced. The problems we have now are problems because the systems are operating at their design limits. To further improve their operating characteristics will require capital improvements projects including systemic renovations and/or system redesign and replacement.

In conclusion, we have come a long way over the past 3 1/2 years, and we have a long way to go. We are very pleased with our six year agreement, and are confident this program will continue to be a winner for both the Board of Education and Honeywell.

In looking back, there is a single ingredient which made and continues to make this project a success—People. Without the commitment of people on both sides, this type of project can never work. If either side approaches this process with the notion that they are going to take all the advantages and not give any to the other side, any such agreement is doomed to fail. Know the people you are doing business with and let them know you. Have a strong, balanced agreement to keep everyone honest and you will both be winners.

Editor's Note

Drawing up Requests for Proposals from scratch, as the Charles County Public Schools (CCPS) did is not necessary. If given the opportunity, CCPS would undoubtedly avoid the creative pain and probably change the RFP they used. The inclusion of the CCPS case study in this text should not be taken to suggest that you use CCPS's RFP as a model document. A basic premise of CCPS's RFP, however, is key; rather than offer a detailed prescriptive RFP, they "decided to allow each respondent to the RFP to create the best package of improvements they could..." Too often, authors of RFPs take false comfort in describing in finite detail exactly what they want. There are two problems with this approach: (1) those issuing the RFP lose the expertise that the proposers could bring to the table—expertise that almost always surpasses that of the organization issuing the RFP; and (2) those issuing the RFP often hire an engineer to get to this level of technical detail, however, they ultimately end up paying the performance contractor again for approximately the same engineering services. (For further discussion refer to Chapter 8,

Establishing Criteria.)

CCPS's procedures are laudable. In particular, the school district first decided what it needed, designed an RFP to meet those needs, drew on the on-staff energy and technical staff as far as possible and then secured outside services as needed. Ideally, however, the consultant should be hired before the RFP is issued. Once the RFP is on the street, you have to live with the good and the bad.

Although the problems vary, the difficulties related to establishing a base year, as discussed in the case study, are not unusual. The sense of open negotiations conveyed in the case study is critical to developing an effective partnership. And that sense of partnership is absolutely essential to an effective project.

The CCPS case study also highlights an area of increasing concern that often emerges during discussions about energy efficiency: indoor air quality (IAQ). While CCPS's indoor air problems preceded any energy efficiency work by the ESCO, the district's recognition of the relationship and the involvement of the ESCO in IAQ mitigation procedures documents the role IAQ will increasingly play in performance contracting.

Finally, the emphasis Lavorgna and Smoyer place on people and communications should be noted; monthly progress meetings, on-site office space for the ESCO project manager and a "tremendous amount of coordination" all speak to this strategically important aspect of performance contracting. Their last paragraph is worth reading a second and third time.

CHAPTER 19

MEETING INSTITUTIONAL NEEDS, AN ESCO PERSPECTIVE

R.Scott Holland

Over the years, climbing energy costs have put a tremendous drain on the public schools' limited resources. When school facilities are old and out of date, structurally unsound or inefficient to operate students' learning opportunities are threatened.

A recent survey by the American Association of School Administrators (AASA), which Honeywell was very pleased to fund, revealed that 74 percent of our nation's schools are living on borrowed time. 31 percent are pre-World War II buildings. Another 43 percent are buildings that were cheaply constructed to meet "baby boom" needs. These energy-inefficient glass boxes were not meant to last more than 30 years.

The AASA survey also revealed that 12 percent, or 1 of 8 school buildings, are inadequate. In the Southeastern states, the problem is even worse, 16 percent, or 1 in 6, are substandard. In an era where we are calling on the public schools to do a better job of educating America's youth, we all need to be concerned about the conditions which are being forced on these students. We are asking teachers to teach and children to learn under conditions where, in many instances, we ourselves would not work.

One crippling factor related to the deterioration of America's schools is the practice of taking money out of maintenance to pay higher utility bills. Deferred maintenance is difficult to measure, but this crushing burden has reportedly now climbed to $100 billion. The costs of deferred maintenance are greater than just deferred costs, for putting off the work accelerates building deterioration, prompts indoor air quality problems, accelerates the need for equipment repair/replacement and contributes to inefficient operations. Ironically, deferred maintenance leads to greater energy consumption, which creates bigger utility bills and that, in turn, drags schools down even further.

Despite the fact that energy costs to the schools have gone up 18 percent in just the last two years, the AASA survey revealed that over half (53.8%) of the nation's school districts do not have an effective energy management program. When asked what prevents them from doing energy work, administrators from every corner of the nation and from every size district have declared overwhelmingly, "a lack of money."

Our public schools face a growing financial and facilities crisis. District budgets have become tighter and tighter every year. At the beginning of 1992, 37 states were currently operating under a budget shortfall. Federal funding was increasingly limited, and local taxpayers were reluctant to fund increases in operating budgets.

Amidst financial binds and deteriorating facilities, performance contracting offers renewed hope. Performance contractors can deliver the resources schools need to cut energy costs roughly 25 percent across the nation. The financing expertise and services of companies like Honeywell can help put the brakes on the decline in the learning environment.

Performance Contracting in Schools

Performance contracting rests on the principal of guaranteed savings. An energy service company will identify cost-effective equipment upgrades/replacement, building modifications and other services to the administration and install those measures that are approved. All the improvements, maintenance on the installed equipment and other enhanced operational procedures are paid for out of future savings. The ESCO guarantees it.

Nearly $2 billion in education dollars could be stopped from going up the smokestack each year and redirect it to the classroom. Performance contracting could make it happen, for school districts pay only if the results are delivered. Payment comes out of savings gained. In this way, performance contracting goes one step beyond funding, as it also guarantees results.

Performance contracting provides an alternative funding base from which schools can solve many of their problems related to aging facilities. The infusion of private sector funds cannot only upgrade or replace equipment, it can make energy efficient modifications to the building shell. The energy savings generated by a performance contract can be used for other school needs.

By going beyond the usual direct purchase procedures, it allows us to demonstrate our confidence in our products and our work. We are willing to risk our resources, betting that our technology and engineering capability will enable school districts to save significant amounts of money. Money that can be utilized to finance new technologies and make needed facility improvements today.

My company, and others like us, are challenged with identifying school districts' operational, financial and technical needs and determining appropriate solutions. Instead of selling a piece of equipment, performance contracting enables us to utilize flexibility in developing a broad range of solutions. Under this approach, we are encouraged to act as providers for a whole spectrum of complimentary facility solutions.

The range of services typically provided in a performance contract include:

- Repair and updating of HVAC equipment
- Maintenance and operation of HVAC equipment
- Computerized energy management systems
- Repair and updating of temperature controls
- Repair and updating of lighting systems
- Repair and updating of fire and security systems
- Indoor air quality diagnostic and mitigating services
- Construction manager for related vendor services
- Customized training programs
- On-site technology resource management
- Facilities management support

Meeting Educational Needs

Balancing comfort and budgets in our schools in not an easy task, however. It requires a business approach that involves overcoming several barriers to implementation.

Not Bid/Spec

The historical bid and specification form of procurement is one of those barriers. And the bid is usually tied to lowest cost. Traditionally, most district's have solicited and awarded building improvement contracts based upon financial criteria alone. Some disadvantages of awarding contracts through this kind of procurement include:

- A bid/spec process, which does not address the quality of service provided. Adherence to bid/spec means vendors are not required to perform to any specific performance standards.
- Lack of vendor evaluation criteria forces districts into a relationship with the lowest bidder, rather than the most effective vendor.
- Short-term, single-service contracts do not meet comprehensive building requirements.

The traditional bidding process simply does not allow for a comprehensive approach that invites cross-disciplinary vendor services, nor does it allow schools to take advantage of long-term operational savings.

To put it another way, when you are in an airplane traveling at 30,000 feet, wouldn't it be scary to think that low bid built that plane? Low bids often end up costing us dearly, but getting across the idea that low bid is not necessarily the answer is not an easy concept to sell.

Overcoming the traditional bidding process requires working with the district's facilities department, administration and board. Convincing them to think differently about solving problems with their aging buildings is key. Getting more flexible procurement procedures is also fundamental.

Performance contracting is a relatively new concept. Selling a contract requires unusual sales techniques and creative strategies. It requires an approach which takes time, because you're asking the district to make a leap of faith. A long selling cycle is commonplace—six to twelve months is average in performance contracting. You're asking the school board to throw their lot in with you for years to come; so they have to be convinced that you will stand behind your claims. In many communities, the school district is the largest employer and the largest budget item. In all communities, the school board and administration are very exposed. Districts must feel confident you can deliver before accepting such risk.

Developing Partnerships

Of course, performance contracting only works when both sides benefit. To be successful, a performance contract must function as a partnership, because either side can ruin the deal through non-performance if the agreement is out of balance. In our work in more than 700 school districts and over 200 college campuses, we have learned that successful partnerships are absolutely essential to success.

From the energy service company (ESCO) perspective, the process associated with forming a partnership begins with:

- an exchange of information,
- an energy and operational audit,
- a review of the district's current programs and objectives for future operation, and
- any comfort or indoor air quality concerns.

This is best accomplished when both sides are open and realistic in their dealings and expectations.

Negotiable Items

Typically, many items will be negotiated. Examples include; operating parameters, room temperatures, lighting hours, light levels, the level of maintenance responsibility, the base period of energy use, etc. Definition of comfort and the learning environment that is acceptable must be the district's responsibility, not the ESCO's.

District personnel need to realize that ESCOs have some items that may not be negotiable. ESCOs cannot negotiate away the right to set specifications for the equipment that may have to be installed under a performance contract. The reason is that the ESCO cannot be placed in the position of guaranteeing performance if it cannot specify what the equipment and systems must accomplish. The installation of this equipment typically falls under the same scenario.

The performance contractor must also retain the right to maintain the system or equipment installed. One U.S. Department of Energy study showed that in any effective energy program, 80 percent of the savings can be attributed to energy efficient operations and maintenance practices. History has shown you can install the best equipment in the world; but if it isn't maintained and operated properly, it will not work any better than the worst piece of equipment you can find. A service company providing a guaranteed performance contract cannot afford to give up the right of maintenance.

There are other contract provisions that are negotiable; the length of the performance contract as well as the close out provisions, for instance. We all need to keep in mind that each item will have some impact on the bottom line. Each has a cost to one side or the other, and each has some impact on the level of savings which can be achieved.

State Legislative Action

All across the country, the case for performance contracting is growing. Many states have enacted comprehensive legislation which promotes performance contracting in schools and government facilities.

One such state is Ohio. It enacted House Bill 264 in 1985. The bill permits Ohio school districts to purchase energy conservation measures on a multi-year installment basis and to increase the portion of a district's net indebtedness that can be used for energy conservation measures. This indebtedness can be incurred without a vote of the people if the funds are used for energy conservation measures that self-fund over a ten year period.

In the first seven years after the legislation was enacted, 130 Ohio school districts have averaged a 126% return on investment. At the current rate of savings, it is estimated that those 130 districts will have saved an average of $57 million per year as a result of performance contracting.

CASE STUDY: OHIO HOUSE BILL 264

The first approved project following the enactment of the law was in April 1986. The benefit of participation is illustrated below as documented through December, 1990.

Year	No. of Districts	No. of Schools	Total Invested
1986	21	132	$23,375,002
1987	25	90	$ 9,764,062
1988	37	193	$21,521,387
1989	39	185	$28,442,084
1990	45	279	$47,771,497
Total	167	879	$131,874,032

Average project cost per school: $150,027
Return on investment through 1990: 126% ROI
Cost avoidance through 1990: **Average $57 million per year**

We All Win

It sounds almost too pat, but performance contracting promotes long-term partnerships where everybody can win. Students and teachers win because classrooms are safer, more comfortable and more conducive to learning.

School districts win because their physical plants are more efficient and less costly to run. Schools will stay that way through preventive maintenance. They also win because they are able to make needed improvements with little risk and without the need for capital expenditure.

Parents win because they know their children are in a productive environment where the focus is on education, as it should be.

And taxpayers win because dollars are not wasted on unnecessary energy, inefficient systems, and too-frequent repairs.

Lastly, companies like Honeywell win because we develop long-term partnerships that return a fair profit to the company.

In a time when the country is calling for education reform and when money to improve facilities is extremely tight, performance contracting is one way we can help relieve the pressure on school districts. Indirectly, we feel we are helping teachers and students, while we build a sound, community-based business. We like that.

CHAPTER 20

MATCHING FEDERAL ENERGY GRANTS WITH PERFORMANCE CONTRACTING

The U.S. Department of Energy (DOE) provides Institutional Conservation Program (ICP) grants to public and private non-profit schools and hospitals to improve the energy efficiency in institutional facilities. These grants require that the institution provide matching funds to implement the project. When schools and hospitals have trouble raising the required matching funds, performance contracting may make it possible. The performance contractor can also help the institution fulfill its reporting and other grant requirements.

For energy service companies (ESCOs), the grants can be incorporated in any agreement that meets the requirements of program regulations. A grant may leverage the project or in other ways make it a more effective way for ESCOs to serve their school and hospital customers.

This chapter offers general guidelines and special considerations for using performance contracting as a match for the federal ICP grants (popularly referred to as the Schools and Hospitals Program). This program is administered through state energy offices, and public administrators may think of it as a state program.

Federal regulations and U.S. Department of Energy (DOE) program directive memorandum (PDMs) influence the implementation of performance contracting as a match for ICP grants. State energy offices also have considerable discretion in operating specific aspects of the program.

If either an institution or an ESCO is considering using a performance contract with ICP, they should consult with the respective state energy office *before* taking any specific action.

A History of Encouragement and Mixed Signals

Since the inception of the Institutional Conservation Program, the only stipulation in the statute regarding local matching funds was that it must be *non-federal* funds. Nothing in the statute has ever prevented the use of performance contracting as a match. In fact, one of the 1987 U.S. Department of Energy Innovation Awards went to the Columbus, Ohio school district, which used performance contracting as a match for an ICP grant.

The DOE has a long history of officially supporting "savings-based arrangements," a term used by DOE's ICP office during the 1980s for all performance contracting agreements. DOE's ICP office also awarded a grant to Hansen Associates in the early 1980s for the project, "Positive Cash Flow Financing in the Institutional Sector," which helped set national guidelines for performance contracting. In 1988, the DOE ICP office published a monograph, *Performance Contracting in the Public Sector.*

The April 1985 federal regulations for ICP (10 CFR Part 455) and *still the governing regulations at this writing* encouraged "savings-based arrangements." The regulation's "Introduction and Description of the Program" stated:

> Savings-based financing agreements provide for the purchase and installation of energy conservation equipment or materials based on the savings produced by the ECM [energy conservation measures]. In general, such agreements do not require an initial capital investment on the part of the institution for the equipment installed under the agreement.
>
> DOE believes that such agreements could aid institutions to reduce their energy use and costs and thus further the goal of institutional energy conservation. In order to maximize the use

of Federal program funds and enhance energy conservation, DOE urges institutions wishing to install conservation improvements to explore all possible funding sources and apply for an ICP grant only if alternative financing is not feasible. [underscore supplied]

Savings-based agreements may also be used as a match for an ICP grant, if the agreements meet the requirements of program regulations. Therefore, DOE is taking this opportunity to explain how the regulations apply to these types of agreements. [underscore supplied]

Savings-based agreements offered to meet the matching or cost sharing requirements of § 455.82(e) may not provide for the retention of title or ownership in any equipment by the supplier or other party beyond the date of project closeout.

In order to avoid conflict of interest, prohibited by the regulations, a savings-based agreement may not provide for a TA analyst to participate in the subsequent installation of the ECM's recommended in the TA. Even though Energy Service Companies will do their own engineering analysis as the basis for the energy service contract, this analysis may not be used as the Technical Assistance Analysis (TA) in the grant application. The TA must be done by an independent analyst.

The regulations require that costs must be established by adequate documentation. In the case of a savings-based agreement compliance with this provision may be achieved by submitting invoices of the supplier or similar documentation.

DOE considers a project ready for closeout under the regulations when the improvements have been made, the equipment has been installed, and all activities specified in the grant award have been completed. Closeout need not be delayed until the end of the period during which the savings are to be shared under a savings-based agreement.

Despite all the encouraging signs noted above, using performance contracting to match an ICP grant has been fraught with confusion and frustration, particularly in certain parts of the country.

During the 1980s, much of the program authority rested with the ten U.S. DOE Support Offices. With regard to performance contracting Support Offices assumed a range of roles from encouragement to passive support. A few were uneasy about the procurement procedures related to performance contracting and tended to shy away from its use. One Support Office opposed performance contracting unless it fit the Augusta College model, which was essentially the use of an RFP to request services to implement *one measure* using a bid/spec low bid "paid from savings" approach. The inconsistency among the Support Offices confused and complicated the situation.

It was complicated even further when DOE, through a PDM, extended its insistence that all institutions be given clear title to the funded equipment upon grant close-out. The intent was apparently to protect DOE's financial interests in the equipment. In effect, this move denied the financial houses the first security interest they needed to protect equipment loans. Some institutions were able to put up other collateral to secure the loan. The net effect, however, was to deny the most needy institutions with their poor credit ratings—the very ones who needed private sector support the most—the use of performance contracting for matching funds.

Congressional Reauthorization and New Regulations in The Making

Against this backdrop, Congress began the process of reauthorizing the ICP. The 1990 law (P.L. 101-440) re-emphasized the original Congressional intent of the law that all *non-federal* funds, including private sector financing, could be used to provide the institution's match. The reauthorization language explicitly provided for the use of performance contracting and also expressly stated that the institution did not have to have clear title to the equipment at grant close-out.

With the passage of the ICP reauthorization language, DOE started its rule-making process to revise the existing ICP regulations to comply with the new law. A Notice of Proposed Rulemaking (NOPR) was issued in the *Federal Register* on January 6, 1992. The comments addressed in the rest of this section of the book are based on U.S. DOE's intentions as expressed in the NOPR and applicable Office of Management and Budget (OMB) circulars. The final rulemaking may vary from the sections quoted below. State energy offices will respond to inquiries regarding regulatory status and provisions.

Under Subpart I—Cost Sharing, of the NOPR (page 456 in the *Federal Register* of January 6, 1992), the following information was provided:

> **§ 455.101 Borrowing the non-Federal share/title to equipment.** The non-Federal share of the costs of acquiring and installing energy conservation measures may be provided by using financing or other forms of borrowed funds, such as those provided by loans and performance contracts, even if such financing does not provide for the grantee to receive clear title to the equipment being financed until after the grant is closed out. However, the grantees in such cases must otherwise meet all the requirements of this part and financing and loan agreements and performance contracts under this section are subject to the requirements of 10 CFR part 600 and the certification requirements under § 455.111(e). Grantees must receive clear title to the equipment when the loan is paid off. [underscore supplied]

The certification requirements under § 455.111(e) stipulate that a clause be inserted in the contract that requires the lender to notify the applicable Support Office Director when a lawsuit is filed seeking a remedy for default.

The next section presented in the NOPR describes the ways that cost-share credit can be incorporated in a project. These provisions may have importance to the grantee and the ESCO.

> **§ 455.102 Energy conservation measure cost-share credit.** To the extent a State provides in its State Plan, DOE may wholly or partially credit the costs of the following, with respect to a building, toward required cost-share for an energy conservation measure grant in that building:
>
> (a) A non-Federally funded technical assistance program dated no earlier than three years prior to the date of the grant application;

(b) A non-Federally funded technical assistance program update to comply with § 455.20(q); and

(c) The non-Federally funded implementation of one or more energy conservation measures, which complies with § 455.71, commenced within the three years prior to the grant application.

Also of interest to grantees and ESCOs is the provisions for using utility rebates from utilities or other entities.

§ 455.104 Rebates from utilities and other entities.

(a) Grantees which receive rebates or other monetary considerations from utilities or other entities for installing the energy conservation measures funded by a grant under this part may use such funds to meet their cost sharing obligations pursuant to § 455.100.

(b) Grantees are not required to deduct the amount of the rebate or monetary consideration from the cost of the measures applied for in a grant application, and DOE does not consider such rebates or monetary considerations to be program income which would have to remitted to DOE upon receipt by the grantee.

Finally, the Davis-Bacon wage rate requirement, which has presented some problems for institutions and ESCOs, is still a provision of the law and the NOPR still sets the threshold at $5,000 for acquisition and installation for any construction contract, and puts it at $2,000 for any subcontract using grant funds.

The conflict of interest provisions, which undergird the regulations, are rooted in OMB circulars. Probably the most striking one that affects performance contracting is the provision that the ESCO cannot have a financial interest in any measures installed with grant funds.

A Step-by-Step Approach

This and other regulatory provisions are reflected in the parallel outline shown on the following pages. This outline was developed to identify the steps that would be taken by a school district in implementing its Institutional Conservation Program (ICP) energy conservation measures (ECM) grant. The outline compares procedures using traditional sources of non-federal funds and performance contracting for the requisite match. *The procedures as outlined were approved by U.S. Department of Energy prior to the legislative reauthorization and regulations related to it.* They also pertain to a specific situation in a particular state. Many of the conditions are tied to provisions in OMB circulars, which have not changed in substance, as they pertain to performance contracting and ICP.

The ICP regulations have and will continue to allow some state discretion; so the first law in matching performance contracting to an ICP grant (PC/ICP) for ESCOs and grantees bears repeating:

CHECK WITH THE STATE ENERGY OFFICE FIRST.

PLEASE NOTE: The procedures here were designed to meet particular concerns in one school district in one state.

It does not address some subtle nuances needed to assure program compliance for other types of institutions or under other conditions in other states. While it takes into account procedures to satisfy program concerns, such as conflict of interest, the reader should not attempt to use it verbatim in another setting. This general approach, therefore, needs to be reviewed with appropriate officials and modified to satisfy DOE Support Office, state and local provisions.

BOARD OF EDUCATION OF ______________________.

ICP PROCEDURES USING TRADITIONAL SOURCE(S) FOR NON-FEDERAL MATCHING FUNDS	ICP PROCEDURES USING SAVINGS-BASED AGREEMENTS FOR NON-FEDERAL MATCHING FUNDS

* *

GRANT RECIPIENT USES OWN PROCUREMENT POLICIES AND PROCEDURES AS SPECIFIED IN OMB CIRCULAR A—110, ATTACHMENT O, "PROCUREMENT STANDARDS, "PARAGRAPH 3 AND ADHERES TO STANDARDS FURTHER SET FORTH IN PARAGRAPHS 3 AND 4 [NOW PART OF THE "COMMON RULE"*], THE SCHOOL ADMINISTRATION CONDUCTS THE FOLLOWING STEPS:

* *

Traditional Sources
No action required

Savings-Based Agreements

1. School district will identify energy and financing needs for its project, maintenance services, training, cost effective energy opportunities not identified in the TA, and a financing structure that assures no negative cash flow.

2. Criteria for selection of a firm will be based on these needs and proven successful practices for performance contracting projects. This criteria will influence the selection of a two step, RFQ/RFP procurement procedure in an effort to maximize competition, maximize opportunities and minimize the risk to the district.

3. The district will issue a Request for Qualifications to all firms known to provide or have interest in providing services and financing to help schools improve energy efficiency. In addition to using an established mailing list, the RFQ will be publicly advertised in accordance with the district's procurement procedures.

4. The proposals responding to the RFQ will be reviewed and assessed as to their ability to meet the previously established energy and financing needs of the district. These firms determined to be capable of meeting the school system criteria and guidelines will be certified as qualified by the school system.

Traditional Sources
No action required

Savings-Based Agreements

5. The school district will notify all firms submitting a proposal as to their qualification status.

6. The second step of the process will be the issuance of a Request for Proposals (RFP) by the district. This RFP will request the firms demonstrate their technical capabilities and associated financial arrangements. The RFP will include summarized specifications of the funded measures without actual grant amounts as required by communique from U.S. Department of Energy. The district realizes that the most accurate and advantageous pricing cannot be obtained until the school system lets bids for the ECM equipment.

7. The district will evaluate the proposals as to:
(a) technical capability
(b) responsiveness to criteria set by the district:
(c) ICP compliance; and
(d) the net financial benefit to the district.

8. The school committee will then recommend to the Board of Education the firm that will best serve its needs. The ESCO will function as a general contractor and will assist in the requisite services and financing.

Traditional Sources
No action required

Savings-Based Agreements

9. The Board of Education will select the firm to serve as the energy service company (ESCO) for the school system and will proceed to negotiate a contract.

****

PROGRAM ADMINISTRATION

****

Traditional Sources

1. District engineering consultant reviews funded measures; including design.

2. District notifies state energy office and obtains approval for any modifications

Savings-Based Agreements

1. ESCO reviews TAs for funded measures; including design. If any ICP funded measures appear to no longer be appropriate, an independent TA analyst, as per ICP regulations, will be retained by the district to review the TAs. This analyst will be free of conflict of interest and will not be paid from the federal ICP grant funds.

2. District notifies state energy office and obtains approval for any modifications.

The district will also consider additional recommendations for equipment, maintenance services, training and any other services that might enhance its energy efficiency. Such considerations, accounting procedures and other matters related to this work will be kept separate from the ICP project.

Traditional Sources	**Savings-Based Agreements**
3. District develops bid specifications for ECM's with assistance of consulting engineer. (In accordance with OMB Circular A-110, eligible bidders may have no financial interest in equipment for which specifications are issued.)	3. The district will develop bid specifications for ECMs with assistance of ESCO. (In accordance with OMB Circular A-110 (applicable to public schools), bid preparers may have no financial interest in equipment for which specifications are issued.)
4. District issues bid requests which assures free and open competition on all funded measures according to OMB Circular A-110, Attachment O, Section 3.	4. District issues bid requests, which assures free and open competition on all funded measures according to OMB Circular A-110, Attachment 0, Section 3.
5. District accepts the lowest qualified bids.	5. District accepts the lowest qualified bids.
6. District relies on its personnel and consulting engineer to oversee installation of measures.	6. District relies on the ESCO with oversight from its own personnel to oversee installation of measures.
7. With consultants, vendors, district inspects measures; accepts measures and title to equipment.	7. With consultants, vendors, district inspects measures; accepts measures and title to equipment.
8. District advises state that measures have been installed in an acceptable manner and requests reimbursement for ECMs.	8. District advises state that measures have been installed in an acceptable manner and requests reimbursement for ECMs.

Traditional Sources	Savings-Based Agreements
9. The State reviews the request for reimbursement, verifies compliance with ICP regulations and guidelines, and reimburses the district up to 50 percent of the actual costs up to the grant amount.	9. The State reviews the request for reimbursement, verifies compliance with ICP regulations and guidelines, and reimburses the district up to 50 percent of the actual costs up to the grant amount.
10. State monitors project, using standard grant monitoring procedures.	10. State monitors project, using standard grant monitoring procedures.

* *

FINANCIAL CONSIDERATIONS

* *

Traditional Sources	Savings-Based Agreements
1. The district assures the state energy office that the nature of its match is non-federal, and no unallowable costs will be ascribed to the ECM portion of the project.	1. The district assures the state energy office that the nature of its match is non-federal, and no unallowable costs will be ascribed to the ECM portion of the project.

*The above guidelines are presented as approved by U.S. DOE in 1989. The author is not aware of any subsequent DOE approved ICP/performance contracting guidance of this nature that would in anyway conflict with the above outline. The designations for OMB Circulars have changed and the designated references no longer apply; however, keeping this material consistent as approved seemed preferable. The DOE ICP can provide current citings. The program regulations were in the change process at this writing and these guidelines should be reviewed in the light of the final 1992 rule-making once published.

Changes on The Horizon

Technical Assistance Auditor

Discussions surrounding the ICP NOPR at this writing suggest changes may be made in the final rule that could make it possible for an ESCO to perform the Technical Assistance (TA) *analysis* (audit).

The key to any effective TA/ECM package under ICP is a high quality TA that identifies the full range of energy conservation opportunities and their interrelation in a building complex. Experience in ICP shows that many TAs made by the "independent analyst" are lacking in completeness, and fail to incorporate recent state of the art technical advances. Insistence on an "independent analyst" does not necessarily result in an objective and complete technical analysis.

If the independent auditor provision were removed, the institution would not have to pay half of the independent TA under the grant program as well as ultimately paying all of a separate audit done by the ESCO. The ESCO would be confident going in that the projects were doable and the auditor was accountable for his/her predictions. The TA would be a true assessment of energy conservation opportunities and not a stepping stone for an ECM application.

Removal of the independent auditor provision was not in the NOPR. Institutions and ESCOs interested in ICP/performance contracting arrangements should contact their state energy offices and ask to be informed when the final rule is issued and whether or not an independent auditor is still required.

State Level Partnerships

A more basic shift in ICP operations at the state level is in the offing. The reauthorizing language and the NOPR outline new opportunities for the state energy offices to join with private sector partners to serve their institutions. Without going into specific legal provisions, the opportunity now exists for states to offer valuable programs to their constituents that have previously been difficult to offer. If a state can show it has the non-federal source(s) available to implement ECMs equivalent to its federal ECM allocation, the state can use all of the federal allocations for TAs, program assistance and marketing.

The innovative opportunities for ESCOs to work with state energy offices in this new relationship are great. ESCOs might provide the federal ECM equivalent in private sector financing for large institutions, where the opportunities for large energy expenditures make performance contracting more attractive; thus, freeing the state to reserve its funds for serving smaller institutions. Initially, those states with revolving funds from oil overcharge money will, of course, be able to use those funds for the nonfederal share if they so chose.

Since the 1970s, the way we have been meeting our energy efficiency needs has been in a state of evolution. The grants program is about to become broader and more flexible. Private sector firms have dramatically changed the way they respond to customer needs. Joint venture opportunities to serve our nation's energy needs are more open than ever. Sharing the excitement of a growing performance contracting industry has been rewarding. Today, performance contracting is poised on the threshold of even greater opportunities to work with government, utilities, institutions and commercial establishments. In the words of Cervantes, echoed by many vaudeville fanciers of *Don Quixote* through the years, "Thou hast seen nothing yet."

REFERENCES AND RESOURCES

REFERENCES

American Council for an Energy-Efficient Economy (ACEEE), 1989. *The Potential for Electricity Conservation in New York State*, American Council for an Energy-Efficient Economy for the New York State Energy Research and Development Authority, Albany, New York (Report #89-12).

ACEEE, 1990 (1). *Lessons Learned: A Review of Utility Experience with Conservation and Load Management Programs for Commercial and Industrial Customers*, American Council for an Energy-Efficient Economy for the New York State Energy Research and Development Authority, Albany, New York (Report #90-8).

ACEEE, 1990 (2). *The Achievable Conservation Potential in New York State from Utility Demand-Side Management Programs*, American Council for an Energy-Efficient Economy for the New York State Energy Research and Development Authority, Albany, New York (Report #90-18).

ACEEE, 1986. *Financing Energy Conservation*, American Council for an Energy-Efficient Economy, Washington, D.C.

Anderson, J., L. Kline, S. Kolbe, J. Raab, A. Rudin and D.R. Wolcott, 1991. "Customer and Trade Ally Backlash to DSM: Point, Counterpoint," in *Proceedings of the Fifth National Demand-Side Management Conference*, Electric Power Research Institute, Palo Alto, California.

Barakat, 1991. *Demand-Side Management Incentive Regulation*, Barakat & Chamberlin, Inc. for the Edison Electric Institute and the Electric Power Research Institute, Washington, D.C.

Bullock, Cary, 1990. "Utilities as Bidders - Competing with ESCOs," in *Proceedings of DSM Bidding: Challenges and Opportunities*, Synergic Resources Corporation, Bala Cynwyd, Pennsylvania.

California Public Utility Commission, 1987. *Economic Analysis of Demand-Side Management Programs*, California Standard Practice Manual, California Public Utility Commission, San Francisco, California.

Center for Environmental Legal Studies, Pace University, 1990. *Environmental Costs of Electricity*, Pace University Center for Environmental Legal Studies for the New York State Energy Research and Development Authority and the U. S. Department of Energy, Oceana Publications, New York.

Cole, W.J., M.J. Weedall and D.R. Wolcott, 1988. "Competitive Bidding of Demand-Side Management," in *Proceedings of the 1988 ACEEE Summer Study on Energy Efficiency in Buildings*, American Council for an Energy-Efficient Economy, Washington, D.C.

Decision Research Corporation (DRC), 1988. *A Qualitative Market Study of the Performance Contracting Program*, Decision Research Corporation for Orange and Rockland Utilities, Pearl River, New York and the New York State Energy Research and Development Authority, Albany, New York.

Edison Electric Institute (EEI), 1990. *State Regulatory Developments in Integrated Resource Planning*, Edison Electric Institute, Washington, D.C.

Electric Power Research Institute (EPRI), 1990. *Impact of Demand- Side Management on Future Customer Electricity Demand*, Electric Power Research Institute, Palo Alto, California (Report #CU-6953).

ERCE, 1991. *Evaluation of the Orange and Rockland Utilities Competitive Bidding Program for Demand-Side Resources*, ERCE Environmental and Energy Services Co. for the New York State Energy Research and Development Authority, Albany, New York.

Esteves, R.M., 1989. "Fact and Fiction in Integrated Bidding Processes: Let's Look at the Record!" *The Electricity Journal*, December.

Fickett, A.P., C.W. Gellings and A.B. Lovins, 1990. "Efficient Use of Electricity," *Scientific American*, September.

Gellings, Clark W. and John H. Chamberlin, *Demand-side Management Concepts and Methods*, Lilburn, GA: The Fairmont Press, 1987.

Goldman, C.A. and J.F. Busch, 1991. "Review of Utility Experience with Demand-Side Bidding Programs," in *Proceedings of the Fifth National Demand-Side Management Conference*, Electric Power Research Institute, Palo Alto, California.

Goldman, C.A. and D.R. Wolcott, 1990. "Demand-Side Bidding: Assessing Current Experience," in *Proceedings of the 1990 ACEEE Summer Study on Energy Efficiency in Buildings*, American Council for an Energy-Efficient Economy, Washington, D.C.

Hansen Associates, 1984. *Positive Cash Flow Financing in the Institutional Sector*, Washington, D.C., U.S. Department of Energy.

Hansen Associates, 1987. *The Bottom Line*. Harrisburg, PA: Pennsylvania Energy Office.

Hansen, Shirley, 1991. *Managing Indoor Air Quality*. Lilburn, GA: The Fairmont Press.

Institutional Conservation Programs, 1988, U.S. Department of Energy. *Performance Contracting in the Public Sector*. Washington, D.C.

King, A., 1990. "What Do Contractors Need and Expect from Utilities," in *Proceedings of DSM Bidding: Challenges and Opportunities*, Synergic Resources Corporation, Bala Cynwyd, Pennsylvania.

Lawrence Berkeley Laboratory (LBL), 1990. *Comparative Analysis of Monetary Estimates of External Environmental Costs Associated with Combustion of Fossil Fuels*, Lawrence Berkeley Laboratory, Berkeley, California (Report #LBL-28313).

LBL, 1991 (1). *Survey of State Regulatory Activities on Least Cost Planning for Gas Utilities*, Lawrence Berkeley Laboratory for the National Association of Regulatory Utility Commissioners, Berkeley, California (Report #LBL-30353).

LBL, 1991 (2). *Independent Evaluation of Niagara Mohawk's Integrated Bidding Program*, Lawrence Berkeley Laboratory for the New York State Energy Research and Development Authority, Albany, New York.

Moskovitz, D., 1989. *Profits and Progress through Least-Cost Planning*, for the National Association of Regulatory Utility Commissioners, Washington, D.C.

Moss, J.B., 1991. "Demand-Side Management: Constituencies, Dissonance and Vendor Roles," annual meeting of the National Association of Energy Service Companies, Washington, D.C.

Nadel, S.M., M.W. Reid and D.R. Wolcott, ed., 1992. *Regulatory Incentives for Demand-Side Management*, American Council for an Energy-Efficient Economy, Washington, D.C. and the New York State Energy Research and Development Authority, Albany, New York.

National Independent Energy Producers, 1990. *Bidding for Power: The Emergence of Competitive Bidding in Electric Generation*, National Independent Energy Producers, Washington, D.C.

New Jersey Board of Public Utilities, 1988. "Stipulation of Settlement in Docket No. 8010-687B," New Jersey Board of Public Utilities, Trenton, New Jersey.

New York Public Service Commission, 1988. "Opinion and Order Establishing Guidelines for Bidding Program," in Case 88-E-241: Orange and Rockland Utilities, New York Public Service Commission, Albany, New York.

Office of Technology Assessment, Congress of the United States. *Energy Efficiency in the Federal Government*, Washington, D.C.: U.S. Government Printing Office, 1991.

Pacific Energy Associates (PEA), 1989. "Responses to Energy Service Company Profile Interviews: Perceptual Questions," Pacific Energy Associates for the New York State Energy Research and Development Authority, Albany, New York.

Peters, J.S., D.J. Barry, M.J. Horowitz and D.R. Wolcott, 1991. "A Dual State Bidding Program: Effects of Different Guidelines," in *Proceedings of the 1991 International Energy Program Evaluation Conference*, U. S. Department of Energy, Washington, D.C.

Peters, J.S. and M.H. Haeri, 1989. *Evaluation of Central Maine Power's Pilot Power Partners Program. Pilot Efficiency Buy-Back Program and Pilot Commercial and Industrial Shared Savings Program*, ERCE Environmental and Energy Services Co. for the Central Maine Power Corporation, Portland, Oregon.

Technical Development Corporation (TDC), 1989. *Detailed Guides to Energy Performance Contracting in New York State's Public Sector*, Technical Development Corporation for the New York State Energy Research and Development Authority, Albany, New York.

Vallen, M.A., 1991. "Overview of the Demand-Side Management Industry," in *Proceedings of DSM: Business and Finance Issues Symposium*, Infocast, Inc., Tarzana, California.

Wolcott, D.R., 1990 (1). "The Pros and Cons of Demand-Side Bidding," Testimony before the *Conservation Report Proceedings*, California Energy Commission, Sacramento, California.

Wolcott, D.R., 1990 (2). "Definition, Evolution and Current Status of DSM Bidding," in *Proceedings of DSM Bidding: Challenges and Opportunities*, Synergic Resources Corporation, Bala Cynwyd, Pennsylvania.

Wolcott, D.R., 1991. "Viability of the Demand-Side Management Industry," in *Proceedings of DSM: Business and Finance Issues Symposium*, Infocast, Inc., Tarzana, California.

Resources

National Association of Energy Service Companies, Ms. Terry Singer, Executive Director, 1440 New York Ave., N.W., Washington, D.C. 20005

State Energy Offices

Alabama Depart. of Economic
and Community Affairs
P.O. Box 5690
Montgomery, AL 36103-5650
(205) 242-5333

Depart. of Community and
Regional Affairs
Rural Development Division
949 East 16th Avenue, Suite 403
Anchorage, AK 99508

Office of the Governor
Territorial Energy Office
Pago Pago, AS 96799

Arizona Depart. of Commerce
Arizona Energy Office
3800 North Central, Suite 1200
Phoenix, AZ 85012

Arkansas Energy Office
One State Capitol Mall
Suite 4E/215
Little Rock, AR 72201

California Energy Commission
1516 9th Street
Sacramento. CA 95814

Colorado Office of Energy
Conservation
1675 Broadway, Suite 1300

Denver, CO 80202
Policy Development and Planning
Office of Policy and Management
80 Washington Street
Hartford, CT 06016

Division of Facilities Management
Energy Office
P.O. Box 1401
Dover, DE 19901

District of Columbia Energy Office
613 G Street, NW, 5th Floor
Washington, DC 20001

Florida Energy Office
Dept. of Community Affairs
2740 Centerview Drive
Tallahassee, FL 32399-2100

Office of Energy Resources
254 Washington Street, SW
Room 401
Atlanta, GA 30334

Guam Energy Office
P.O. Box 2950
Agana, GU 96910

Depart. of Planning and
Economic Development
Energy Division
335 Merchant Street, Room 110
Honolulu, HI 96813

Depart. of Water Resources
Energy Division
1301 North Orchard
Boise, ID 83706

Depart. of Energy and
Natural Resources
325 West Adams Street, Room 300
Springfield, IL 62704-1892

Depart. of Commerce
Office of Energy Policy
1 North Capitol, Suite 700
Indianapolis, In 46204-2248

Iowa Energy Bureau
Transportation and Planning Section
Wallace State Office Building
Des Moines, IA 50319

Kansas Corporation Commission
Energy Programs Section
1500 SW Arrowhead Road
Topeka, KS 65504

Kentucky Division of Energy
691 Teton Trail
Frankfort, KY 40601

Louisiana Depart of Natural
Resources
Energy Division
P.O. Box 44156
Baton Rouge, LA 70804-4156

Depart. of Economic and
Community Development
219 Capital Street
Augusta, ME 04333

Maryland Energy Administration
45 Calvert Street
Annapolis, MD 21401

Massachusetts Division of
Energy Resources
Leverett Saltonstall Building
100 Cambridge Street, Room 1500
Boston, MA 02202

Michigan Public Service Commission
P.O. Box 30221
6545 Mercantile Way
Lansing, MI 48909

Minnesota Depart. of Energy and
Economic Development
Department of Public Service
900 American Center Building
150 East Kellogg Boulevard
St. Paul, MN 55101

Mississippi Depart. of Economic
and Community Development
Energy and Transportation Division
510 George Street, Suite 101
Jackson, MS 39202-3095

Missouri Depart. of Natural
Resources
Division of Energy
P.O. Box 176
Jefferson City, MO 65102

Depart. of Natural Resources
and Conservation
Energy Division
1520 East Sixth Avenue
Helena, MT 59620-2301

Nebraska Energy Office
P.O. Box 95085
State Capitol Building, 9th Floor
Lincoln, NE 68509

Nevada Office of Community
Services
Capitol Complex
Carson City, NV 89701

Governor's Office of Energy
and Community Services
57 Regional Drive
Concord, NH 03301-8506

Dept. of Environmental
Protection and Energy
401 E. State Street
Trenton, NJ 08625

Depart. of Energy, Minerals
and Natural Resources
Energy Conservation and
Management Division
2040 South Pachoco
Santa Fe, NM 87505

New York State Energy Office
Division of Conservation
2 Rockefeller Plaza
Albany, NY 12223

North Carolina Depart. of
Economic and Community
Development
Energy Division
430 North Salisbury Street
Raleigh, NC 27611

Office of Intergovernmental
Assistance
State Capitol Building
Bismarck, ND 58505

Commonwealth Energy Office
P.O. Box 340
Saipan, CM 96950

Ohio Depart. of Development
Community Development Div.
Office of Energy Conservation
77 S. High Street, 24th Floor
Columbus, OH 43266-0413

Oklahoma Depart. of Commerce
Division of Community Affairs
and Development
P.O. Box 26980
Oklahoma City, OK 73126-0980

Oregon Depart. of Energy
Conservation Division
625 Marion Street, NE
Salem, OR 97310-0831

Pennsylvania Energy Office
116 Pine Street
Harrisburg, PA 17101-1227

Dept. of Consumer Affairs
P.O. Box 41059
Minillas Station
Santurce, PR 00940

Governor's Office of Housing,
Energy and Intergovernmental
Relations
State House
275 Westminster Street, Room 143
Providence, RI 02903-5872

Office of the Governor
Division of Finance and
Administration
Office of Energy Programs
1205 Pendleton Street 3rd Floor
Columbia, SC 29201

Governor's Office of Energy Policy
217 1/2 West Missouri
Pierre, SD 57501

Tennessee Depart. of Economic
and Community Development
Energy Division
320 6th Avenue North, 6th Floor
Nashville, TN 37219-5308

Governor's Energy Office
Capitol Station, P.O. Box 12428
Austin, TX 78711

Utah Energy Office
355 West North Temple
3 Triad Center, Suite 450
Salt Lake City, UT 84180-1204

Department of Public Services
Energy Efficiency Division
120 State Street
Montpelier, VT 05620

Virgin Islands Energy Office
81 Castle Coakley
Christiansted
St. Croix, VI 00823

Depart. of Mines, Minerals
and Energy
2201 West Broad Street
Richmond, VA 23220

Washington State Energy Office
Conservation Division
809 Legion Way, SE
Olympia, WA 98504

Governor's Office of
Community and Industrial
Development
Fuel and Energy Office
State Capitol Complex
Building 6, Room 553
Charleston, WV 25305

Wisconsin Division of Energy
and Intergovernmental
Relations
101 South Webster Street
P.O. Box 7868
Madison, WI 53707-7868

Division of Economic and
Community Development
Energy Division
Herschler Building, 2nd Floor
Cheyenne, WY 82002

ANSWERS FOR WORKSHEET 5-1, PAGE 44

CALCULATING ENERGY CONSUMPTION

Building ____________________ Period __________ to __________

(month or year)

Gross square feet 47,000

	UNITS CONSUMED		BTU/UNIT		MILLION BTU (MMBTU)
Electricity	1,417,000	kWh ×	3,413	=	4,836.2 MMBTU
Natural gas	930	mcf ×	1,030,000	=	957.9 MMBTU
Distillate oil		gal ×	138,690	=	
Residual oil		gal ×	149,690	=	
Other		×			
			TOTAL		5,794.1 MMBTU

Divided by 47,000 sq. ft. = 123,279BTU/SQ.FT.

GLOSSARY OF TERMS

BASE YEAR (Baseline consumption)—A recent year, or average of years, used as a reference base to compute savings attributable to the energy conservation measures being financed.

COST-BASE (applied to savings-based agreements)—The total cost of the equipment and related services provided to the organization plus a percentage fee to cover energy service company's operating margin and indirect costs.

COST-BASED CONTRACT (or paid from savings contract)—A variable term based financing agreement in which a cap is placed on the total amount of payments to the financing firm. The cap is reached when the present value of all cumulative payments to the firm equals the cost-base plus a pre-determined profit percentage.

ENERGY SERVICE COMPANY—A firm which provides energy management services including an engineering evaluation of the building, financing and installation of energy-saving equipment and procedures; and provides an agreed upon comfort level for a fee usually guaranteed not to be exceed the building's avoided energy costs.

ENERGY SERVICE PROVIDER—An energy service company or registered professionals, such as architectural and engineering firms, that provide financing and/or expertise for a share of the avoided costs.

GUARANTEED SAVINGS—Agreement, or contract clause, whereby a savings-based financing firm will guarantee that a piece of equipment, or package of energy conservation measures, will achieve a minimum amount of savings over a contract period, or the firm will refund the difference between the actual and guaranteed savings (or otherwise compensate the organization).

HURDLE RATE—The point at which it becomes financially beneficial to the organization to use outside funds: usually based on present rate of interest earned on internal funds and related conditions.

INSTALLMENT/PURCHASE—An arrangement whereby the purchaser makes payments at regular intervals until the cost of the equipment has been satisfied. Unlike lease/purchase, ownership is predetermined and the contract may exceed a fiscal year without automatic renewal provisions. (Sometimes used interchangeably with the term "municipal leasing.")

INTERNAL FINANCING—Financing with money available or retained by the organization without securing outside revenues.

INTERNAL RATE OF RETURN—The discount rate that discounts an investment's expected net cash flows to a net present value of zero.

LEASE—An agreement to make regular payments to a lessor (owner) over a set period of time in exchange for the use of a building or equipment.

LEASE PURCHASE (closed-end lease)—An arrangement whereby a lessee commits to making payments for the use of buildings or equipment for a set period at which time the lessee has the right to buy the property for a price agreed upon in advance, frequently a nominal figure.

LESSEE—The user of a leased asset who pays the lessor for the usage right.

LESSOR—The owner of a leased asset.

MINIMUM ACCEPTABLE RATE OF RETURN—The lowest rate of return that an investment can be expected to earn and still be acceptable; same as the investment's cost of capital.

NET PRESENT VALUE (NPV) (of energy savings)—The value in today's dollars (year zero) of future energy savings less all project contracting, financing, and operating costs. This measure takes into account the "time value of money."

OPEN-ENDED LEASE—A lease where there is no fixed price purchase option at the conclusion of the lease period. (Lease may indicate "purchase at fair market value" but no dollar amount is specified.)

OPERATING LEASE—A lease that is cancellable by the lessee at any time upon due notice to the lessor. Also refers to a short-term lease that is cancellable by the lessor or lessee upon due notice to the other party.

PAYBACK PERIOD—The amount of time required for an asset to generate enough net positive cash flow to just cover the initial outlay for that asset.

PERFORMANCE CONTRACT—A contract with payment based on performance; usually a guarantee that costs will not exceed energy savings. (Also see savings-based agreement.)

POSITIVE CASH FLOW LEASE—A closed-end lease (lease/purchase) in which payments made to the lessor are kept below the level of savings derived from the leased equipment (for each leased payment, a specified time period or the total lease period.)

PRESENT VALUE—The value of money at a given date (current or today's dollars) that will be paid or received in future periods.

PRIME RATE—Interest rate charged by banks on short-term loans to large low-risk businesses.

PROJECTED SAVINGS (in a savings-based financing agreement)—Refers to the expected annual dollar value of the reduced energy consumption due to implementing conservation measures.

RATE OF RETURN—The interest rate earned on an investment; may be the actual rate or expected rate.

SALE AND LEASEBACK—An arrangement under which the user of the asset sells the asset and then leases it back from the purchaser.

SALVAGE VALUE (residual value)—The money that an organization can receive for an asset after it has held it for a period of time.

SAVINGS-BASED AGREEMENT (OR CONTRACT)—Arrangements by which energy service companies agree to finance and maintain energy efficiency improvements in a client's buildings with the repayments to come from an agreed upon percentage of the savings generated by the improvements. A generic term used by the U.S. Department of Energy for any financing agreement with payment predicted on energy savings.

SAVINGS-BASED FORMULA—The formula (calculation of savings procedure) specified in the contract which is used to determine savings. Usually involves four steps: (1) determine base year usage; (2) adjust base year usage for variations (temperature, occupancy, etc.); (3) subtract actual usage from adjusted base year consumption; and (4) calculate savings by multiplying the units of energy saved by the current cost per unit. Calculations for electrical demand savings are considered part of the formula but computed separately.

SENSITIVITY ANALYSIS—Analysis of the effect on a project's cash flows (or profitability) of possible changes in factors which affect the project; e.g., level of predicted savings, energy price escalation, etc.

GLOSSARY OF ACRONYMS AND ABBREVIATIONS

A&E	architectural and engineering services, or firm
AASA	American Association of School Administrators
ACEEE	American Council for an Energy-Efficient Economy
AEE	Association of Energy Engineers
APP	adjusted payback period
ASHRAE	American Society of Heating, Refrigerating and Air-Conditioning Engineers
BAFO (B&F)	best and final offer
Btu	British thermal unit
CBO	Congressional Budget Office
CFC	chlorofluoro carbon
cfm	cubic feet per minute
COBRA	Comprehensive Omnibus Budget Reconciliation Act of 1985
COD	cost of delay
DD	degree days
DOD	U.S. Department of Defense
DOE	U.S. Department of Energy
DSM	demand-side management
ECI	energy cost index
ECM	energy conservation measure
EEI	Edison Electric Institute
EMS	energy management system; or energy management control system
EPA	U.S. Environmental Protection Agency
EPRI	Electric Power Research Institute
ERAM	electric rate adjustment mechanism

ESCO	energy service company
EUI	energy utilization index
FAR	Federal Acquisition Regulations
FEMIA	Federal Energy Management Improvement Act of 1988
FMS	facilities management system
HBI	Healthy Buildings International, Inc.
HUD	U.S. Department of Housing and Urban Development
HVAC	heating ventilating and air-conditioning system
IAQ	indoor air quality
ICP	Institutional Conservation Program, U.S. Department of Energy; a 50/50 matching grants program for schools and hospitals
IRP	integrated resource planning
kW	kilowatt
kWh	kilowatt hours
LBL	Lawrence Berkeley Laboratory
LCC	life-cycle costing
MCF	thousand cubic feet
MMBtu	million Btu
NAESCO	National Association of Energy Service Companies
NDAA	National Defense Authorization Acts
NIOSH	National Institute of Occupational Safety & Health, U.S. Department of Health & Human Services
NOAA	National Oceanic and Atmospheric Administration
O&M	operations and maintenance
OTA	Office of Technology Assessment, Congress of the United States
PEPCO	Potomac Electric Power Company
PM	preventive maintenance
PSC	public service commission
PUC	public utility commission
PURPA	Public Utility Regulatory Policies Act
RFP	request for proposals
RFQ	request for qualifications
ROI	return on investment
SBS	sick building syndrome
SCF	simplified cash flow
SES	shared energy savings; used by the federal government

SPP	simple payback period
TFC	termination for convenience
TRC	total resource cost
VOC	volatile organic compound

APPENDIX A

REQUEST FOR QUALIFICATIONS MODEL LANGUAGE

The request for qualifications (RFQ) appearing on the following pages offers suggested language for a public institution interested in securing energy services and financing. The RFQ language presented in this appendix does not purport to cover all conditions or eventualities that may arise. It is intended only as a model document that must be modified to meet specific concerns, circumstances and conditions unique to a given facility or complex.

BOARD OF EDUCATION OF ALPHA BETA CITY
REQUEST FOR PROPOSALS: ENERGY SERVICES

SCHEDULE OF EVENTS

REQUEST FOR PROPOSALS	XXXXXX, 1992
PRE-PROPOSAL CONFERENCE:	Date: XXXXXXX, 1988 Time: 2:00 p.m.

LOCATION: Board of Education
Alpha Beta City, Maryland 21000

PROPOSAL DEADLINE:	XXXXXXX, 1992 2:00 p.m. local time

SUBMIT TO: Mr. Administrator,
Director of Purchasing
Board of Education
123 Main Street
Alpha Beta City, Maryland 21000

PROPOSAL OPENING:	XXXXXXX, 1992
SELECTION OF FIRM:	XXXXXXX, 1992

BOARD OF EDUCATION OF ALPHA BETA CITY
REQUEST FOR QUALIFICATIONS: ENERGY SERVICES

I. PURPOSE AND SCOPE

The Board of Education of Alpha Beta City (hereinafter referred to as the "Board" or "Owner") and by reference the Alpha Beta City school system (hereinafter referred to as "ABC") is a public school system interested in receiving proposals from qualified energy services companies (hereinafter referred to as ESCO, firm, proposer, contractor) for providing comprehensive energy management services.

Proposals are requested from firms capable of providing equipment and/or services necessary to achieve cost-effective energy efficiency, reduce the district's operating costs and to serve other district facility needs.

Evaluation procedures will not only consider the experience of the firm and its assigned personnel, but will weigh the technical feasibility, economic viability, soundness of project financing arrangements, and the estimated net financial benefit to ABC. The district intends to award a negotiated contract to one firm to provide the equipment and services under terms and conditions ABC considers to be most favorable among those offered.

B. SCOPE OF WORK

The company selected as a result of the combined qualification and technical evaluation will be expected to:

1. Provide comprehensive energy management services for all ABC facilities, including but not limited to:

    - a comprehensive energy audit;
    - the design, selection and installation of equipment systems, and modifications to improve energy efficiency without sacrificing comfort or existing equipment performance or reliability;
    - the training of district's operations and maintenance personnel in energy efficient practices;

- the maintenance and service of the installed measures;
- 24 hour monitoring and associated field support; and
- financing for the transaction.

2. Assure that any payments for energy efficient improvements or services related thereto are contingent on energy savings so that the Board will not have any financial obligation that exceeds the district's share of t.he avoided utility costs. Explicitly state any costs or obligations incumbent on the board that will not be covered by guaranteed savings.

3. ABC is also interested in capital improvement work, which may not meet the standards for cost-effective energy conservation measures. The district does not wish to incur any negative cash flow in order to accomplish this work. The proposer is asked to suggest any additional measures that could be performed that fits these conditions.

II. FACILITIES

The ABC buildings for which energy efficiency services are requested are listed in Attachment A. These facilities occupy approximately XXXXXX square feet and have an annual utility bill of $XXXXXX. Attachment A offers pertinent utility data and information on the facilities in order to provide the ESCO information upon which to judge the economic viability of the project.

ESCOs will be asked to consider all potentially cost-effective energy efficiency improvements in these facilities.

III. RFQ PROCEDURES

A. ISSUANCE, CONTRACTS, QUERIES

This RFQ is issued by the Board of Education of Alpha Beta City. The following persons should be contacted for additional information regarding this RFQ.

Alpha Beta City school system administration, governance, or financing:

____

____

____

ABC physical plant and/or energy data:

____

____

____

Proposal procedures:

[District personnel or consultant if one is involved.]

Verbal communication will be offered as a matter of clarification; however, such communication by employees or representatives of ABC school district concerning the RFQ shall not be binding on the district and shall in no way excuse the competitor of obligations as set forth in this RFQ.

Any inquiries received in writing on or before eight (8) working days prior to the deadline noted in Section III. D., wherein a response is deemed valuable to the process, the question(s) and response(s) will be sent to all ESCOs of record.

B. <u>SUBMISSION REQUIREMENTS</u>

Proposers should submit an original and eight (8) copies of their proposal. The proposal should be limited in length to thirty (30) pages, exclusive of the sample contract, Appendices and Attachments as described in Section V,

A letter of transmittal must be signed by an official with authority to bind the proposer contractually. The name and title of the individual signing the proposal should be typed immediately below the signature.

C. PREPARATION OF PROPOSALS

Proposals must be in correct format and complete. Elaborate proposals and brochures are not desired. Clarity and concise, orderly treatment are important. ESCOs must address each item in order in which it appears in Section V, "Proposal Format, Content and Specific Criteria," of this RFQ and note the appropriate section heading being addressed at the top of the respective page.

The Project Summary Sheet must be completed . The ESCO is expected to respond to all items in as much detail as necessary for ABC administration, its representatives and consultants to make a fair evaluation of the ESCO and the proposal for ranking. ESCOs should respond directly to the points raised as concisely as possible.

D. DEADLINE FOR PROPOSALS

Proposals must be received on or before 2:00 p.m., XXXXXXX, 1992 at the following address:

ATTN: ____________________

____

____

The Board reserves the right to disqualify proposals received after the time and date specified.

Proposals and all conditions therein shall remain in effect for at least ninety (90) days from proposal submission date.

Proposals, which are incomplete, not properly endorsed, do not follow the requested format, or otherwise are contrary to the guidelines of this RFQ, may be rejected as nonresponsive.

E. DISPOSITION OF PROPOSALS

All proposals become the property of the Board and will be returned only at the owner's option and the proposer's expense. In any event, one copy of each proposal will be retained for the institution's official files.

F. PROPRIETARY DATA

If a proposal includes any proprietary data or information that the respondent does not want disclosed to the public, such data or information must be specifically identified and will be used by ABC administration, and their consultants solely for the purposes of evaluating proposals and conducting contract negotiations.

All proposals, exclusive of supplemental sheets designated "Proprietary Data," will become a matter of public record on the deadline date stipulated in Section III, D.

Each ESCO agrees by submitting a proposal that ABC administration has the right to use any or all ideas or concepts presented in any proposal without restriction and without compensation to proposer thereof, except for specific proprietary data as provided in this section.

G. MODIFICATION OR WITHDRAWAL OF PROPOSALS

Any proposal may be withdrawn or modified by written request of the proposer provided such request is received by the Board by the deadline and at the address as stipulated in Section III, D. Modifications received after the due time and date will not be allowed.

The district or its representatives reserves the right at any time to request clarification from any or all contractors submitting a proposal.

H. PRE-PROPOSAL CONFERENCE

On XXXXXXX, 1992 at 2:00 p.m., the Board of Education of Alpha Beta City will hold a Pre-Proposal Conference in the following location:

____

____

____

The ABC school administration, representatives and its consultant will be present to answer questions regarding the RFQ procedures and the overall project. Floor plans, available as-builts and other pertinent information will be made available at the conference. Those attending the conference will be given an opportunity to walk through three representative facilities.

Attendance is not required and will not be a factor in evaluating proposals. Printed information provided at the conference will be mailed to prospective proposers on request.

ESCOs interested in attending the pre-proposal conference should contact ________ at (___) ____ - _______ [phone number] no later than XXXXXX, 1992.

I. RIGHT TO REJECT

This RFQ does not commit the Board to award a contract or to pay costs incurred in preparation of a bid in response to this request. The school district reserves the right to accept or reject in part or in its entirety, any bid as a result of this RFQ.

J. COST OF PROPOSAL PREPARATION

The cost of preparing a response to this RFQ is not reimbursable to ESCOs or the selected contractor.

K. CONTRACT REQUIREMENTS

Standard construction contract provisions, such as OSHA laws, conditions of site, acceptance, permits, inspection provisions, co-operation with site personnel, indemnity, liability insurance, workmen's compensation, blanket fidelity and the like will be part of the contract terms.

The following conditions will also be incorporated in the contract.

1. Taxes

The district will forthwith pay all taxes lawfully imposed upon it with respect to the equipment. The ESCO will forthwith pay all taxes imposed upon it with respect to the equipment.

By this section, the district makes no representation whatsoever as to the liability or exemption from liability of the ESCO to any tax imposed by any government entity.

2. Choice of Law

The agreement shall be governed in all respects by the laws of the State of Maryland.

3. Performance Bond

The Board may require the submission of a performance bond by the successful proposer upon agreement of the measures to be implemented, up to eighty percent (80%) of the listed purchase price of the equipment. Such bond would then remain in effect and be subject to forfeiture for non-performance until the equipment is accepted by the district.

4. System Integration

A proposal will not be acceptable unless the proposer expressly indicates an intention to be responsible for systems integration, testing and maintenance of all installed hardware and systems level software. Integration of newly installed equipment and controls with existing equipment is considered a critical aspect of implementation.

5. Negotiations

The district reserves the right to negotiate with the successful proposer any additional terms and conditions, which may be necessary or appropriate to the accomplishment of the purpose and scope of the RFQ.

6. Sample Contract

ESCOs are requested to attach to their proposals a recently executed contract with a non-profit institution, preferably a public school, for work similar to that requested herein.

7. Upgrade Assumptions

It should not be assumed when developing the proposal that the district will be able to upgrade or make modifications to existing building equipment. The contractor should include in his proposal all costs to modify, upgrade or fine-tune any building equipment which the contractor feels is necessary to achieve the projected savings.

L. REPRESENTATIVE ENERGY AUDIT

Under separate cover, the proposer will submit an energy audit which the ESCO has performed on a facility(ies) similar to ABC facilities. By its submission, the proposer attests that the sample energy audit is representative of the comprehensiveness, technological sophistication, formulas, calculations and detail it proposes to use under this scope of work.

Each ESCO must state that its contract with the Board will contain the information specified in this section.

M. PROPOSAL EVALUATION AND SELECTION PROCEDURES

All proposals will be evaluated by an ABC district committee and its representatives and consultants. The proposals will be evaluated according to the criteria listed in Section IV and delineated further in Section V.

The evaluation committee will make a recommendation to the Board for their approval. Upon approval, a letter of intent will be sent to the selected ESCO. The selected firm and those representing Board will then seek to negotiate a satisfactory contract within sixty (60) days. If the parties fail to agree on terms of a contract within the (60) day period, the Board reserves the right to terminate all negotiations and either select one of the other finalists or issue a new RFQ.

IV. SELECTION CRITERIA

Evaluation will be made according to the following weighted criteria:

	CRITERIA	WEIGHTING
A.	Proposal presentation	5
•	adherence to format requirements	
•	completeness and clarity	
B.	Proposer's Qualifications	40
•	management plan	
•	experience in similar facilities and demonstrated ability	
•	personnel qualifications and availability for project implementation	
C.	Technical Quality and Range of Services	20
•	technical approach	
•	comprehensiveness of approach	
•	baseline methodology and calculations	
D.	Ability to Implement Project Properly	10
•	management	
•	quality assurance	
•	schedule of implementation	
E.	Financial Considerations	25
•	net financial benefit to district	
•	savings calculations; previous experience in meeting predicted savings	

V. FORMAT, CONTENT AND SPECIFIC CRITERIA

Proposers will follow the format and content outline presented below and each section must be clearly labeled as to the section being addressed. Numbers on specific criterion are used for organizational purposes only and should not be taken as indicating any relative importance of that criterion.

PROJECT SUMMARY SHEET

A. PROPOSAL PRESENTATION:

Criteria:

1. clarity, conciseness and completeness of proposal presentation; and
2. responsiveness to RFP requirements.

B. PROPOSER'S QUALIFICATIONS:

Management Plan. Describe the proposed organization briefly, noting management responsibility for the following areas:

1. Financing
2. Engineering design
3. Construction and/or installation
4. Operation and maintenance; training
5. Other major activities

Qualifications. Describe your ESCO's: (1) corporate capabilities; (2) number of years firm has been involved in providing energy services and number of energy service contracts entered into; (3) years the firm has been involved in providing energy services to schools; and (4) the financial condition of the firm, including a 10 K for publicly held firms, or an annual report.

Also provide information on methods, techniques and equipment used by the ESCO in such matters as design preparation, systems installation, on-site coordination, system operation and monitoring staff training, maintenance and the like.

Demonstrated ability. Provide a brief description of the five most significant recent or current projects of a similar nature conducted by your firm. Include documentation of project approach, management, technical configurations, arrangements, energy savings projections and results achieved, and other pertinent information. Present the information on each project as follows:

1. Client (name, address)
2. Client contact (name, title, phone number)
3. Project title
4. Project description
5. Energy and cost savings projected/achieved and client's share
6. Proposed schedule (from________ to __________) and implementation achieved (________to__________)
7. Total project cost $ and proposer's share $

Personnel assigned. Indicate the personnel that will be assigned to this project and their specific project responsibilities. Indicate their qualifications to assume the assigned responsibilities, including degrees, special training, licenses, years of experience, and special areas of expertise that will enable them to meet these responsibilities effectively. Indicate the percentage of time each individual will devote to the ABC project during its initial phase. A full resume for each assigned person should be included in Appendix B of the proposal.

Criteria:

1. Firm's background and experience; including direct experience in performance contracting with school districts; quality of references;
2. Documentation of ESCO's financial condition and stability;
3. Proposer's experience and reputation in similar projects to that described in this RFP;
4. Assigned personnel's qualifications; project availability
5. Overall ability to maximize benefits and minimize risks to the institution.

C. TECHNICAL APPROACH AND RANGE OF SERVICES OFFERED

Baseline Methodology. Describe the methodology typically used by the firm to compute the energy baseline. Attach a sample computation from a previous job done by your firm with sufficient documentation of methods, assumptions, and input data so that evaluators can follow the procedures used.

Range of services. Describe the complete range of energy services being offered by your firm; e.g., auditing, equipment selections and installation, operations and maintenance, etc. Discuss the mechanism proposed, which will guarantee the local support services necessary for fulfilling the contract terms. Include any contemplated changes to the physical parameters in the facilities; e.g., temperature, lighting levels, humidity and ventilation.

Specify any exceptions or waivers to the contemplated scope of work described herein.

Criteria:

1. Quality of technical analysis and recommendations;
2. Adequacy of equipment or services to accomplish scope of work;
3. Method of selection of energy efficiency equipment;
4. Monitoring of energy usage;
5. Technical field support; emergency response provisions;
6. Reporting procedures;
7. Maintenance on installed equipment; on any additional equipment;
8. Training; and
9. Other services.

D. ABILITY TO IMPLEMENT PROJECT PROPERLY

Management. Describe the specific responsibilities, line of communications, and authority of the management structure. Note interfaces with the ABC district personnel. Indicate typical procedures for identifying problems and preventing schedule slippages and cost overruns.

Quality assurance. Provide a plan for assuring quality of workmanship and provision of services. The plan should indicate who will review work in progress and work products, the schedule for reviews, how top management will be kept informed, and how corrective action will be identified, implemented and reviewed.

Implementation schedule. The proposer's projected implementation schedule of the tasks and responsibilities outlined in the proposal should be included in the proposal as part of this subsection. Any time required to secure financing, start up date, and rate of installation should be included.

Criteria:

1. Plan to provide effective management procedures and quality assurance;
2. Required time frame for ESCO to obtain financing and initiate the project; and
3. Implementation schedule.

E. FINANCIAL CONSIDERATIONS

Financial arrangements. Describe the financial arrangements you propose to fund the cost-effective energy conservation measures you recommend. If an outside funding source is used, discuss the general nature of investors with which you have had experience and the procedures you would expect to pursue.

Note any previous experience with financial arrangements of this type if not covered Section V. B. Also note any tax or legal issues or uncertainties known to you which might affect ABC's cost savings.

Savings calculations. Describe the procedure that will be used to measure actual energy cost savings and the value of such savings attributable to the contracted services. In addition to calculations relative to quantity consumed, include procedures to share savings related to demand and time of day cost adjustments.

Baseline adjustments. Include an explanation of how the savings calculation will be adjusted to reflect changes in weather, occupancy and use; e.g., addition or removal of energy-consuming equipment, changes in the hours or level of occupancy, etc.

Share of savings. Propose a method of allocating the value of the energy savings between ESCO and the institution. Clearly state the number of years that savings will be shared and the proportional allocations of savings each year. Explain any assumptions used.

Sample calculations. Present sample calculations for a seven-year project with $2,500,000 equipment acquisition and installation costs (not total project cost) with a projected payback of three years. Include maintenance on the installed equipment you list and other proposed services. Give estimated total cost. (Label Option A.)

Using the same parameters set forth in the above paragraph, describe the financial arrangements that might be used to provide the capital improvement items mentioned in Section I, B, 3. Give estimated total cost for your example. (Label sample calculations Option B.)

Buyout, termination. and contract continuation. Explain the conditions under which this contract might be continued, extended, or amended beyond the contract years. Describe how the equipment and servicing responsibilities will be affected at the conclusion of the contract. Should it be a factor in the suggested financing approach, describe how the value of the equipment will be calculated upon contract expiration. Provide a;buy-out schedule for Option A and Option B.

In the event of contract termination, describe provisions for assuring that all affected district facilities will be restored to conditions at contract origination or better related to any work rendered under contract.

Contract conditions. State the firm's willingness to comply with the contract requirements discussed in Section III. Note any exceptions to this compliance and the rationale for exclusion. Submit a sample contract which was recently executed with a similar institution.

Criteria:

1. Potential net economic benefit as evidenced in sample material and calculations;
2. Innovative energy financing procedures;
3. Liability and casualty tax insurance on the equipment and personnel involved in providing the service;
4. Source of financing;
5. Protection against poor performance;
6. Calculation of savings—procedures for addressing such variables as changes in building use, utility rates, weather fluctuations;
7. How the baseline energy consumption data is derived;
8. Value of extended services;
9. Methods used to determine amount of ESCO's compensation;
10. Typical termination and buyout provisions;
11. How equipment is valued at end of contract; and
12. Contract renewal options.

APPENDIX A. RESUMES OF ASSIGNED PERSONNEL (REQUIRED)

APPENDIX B. SAMPLE CONTRACT (REQUIRED)

APPENDIX C. ANY EXTENDED DESCRIPTION OF FIRM'S QUALIFICATION AND EXPERIENCE GERMANE TO THE PROPOSED WORK (OPTIONAL)

ATTACHMENTS

SUPPLEMENT: SAMPLE ENERGY AUDIT (REQUIRED)
PROPRIETARY INFORMATION (OPTIONAL)

REQUEST FOR PROPOSALS

The difference between a RFQ and a request for proposals (RFP) procedure usually rests on a two phase selection process. The two phase procedure is generally designed to pre-qualify bidders so that more exacting requirements are only imposed on a short list of bidders. A RFP procedure is not recommended unless the project is very large, very complex, or unique.

The RFQ presented above can be converted to a request for proposals (RFP) by adding reference to a two phase procedure and instruct the potential proposer's as to the purpose of the two phase process; i.e., a test audit. This information will help the ESCO decide whether or not to bid. Language eliciting technical competence should set the auditing and financial parameters.

Should the administration decide they wish to have a RFP procedure in order to assess the final bidders' abilities to approach unique problems through a test audit on given portion of the facilities, language similar to the following could be inserted in the above document. The language shown in capital letters is model language; lower case is explanatory language.

Under III, B of the RFQ insert the following:

TWO PHASE PROCEDURE:

THIS REQUEST FOR QUALIFICATIONS (RFQ) CONSTITUTES THE FIRST PHASE OF A TWO PHASE SELECTION PROCEDURE. THE INTENT OF PHASE 1 IS TO PRE-QUALIFY BIDDERS FOR THE MORE TECHNICAL REQUEST FOR PROPOSAL BY SELECTING _____[A number may be specified if the intent is to narrow the number of firms for phase 2] ENERGY SERVICE COMPANIES, WHICH APPEAR TO BE MOST QUALIFIED TO DO THE WORK AS REQUESTED IN THIS RFQ.

SUCCESSFUL PROPOSERS FOR THE FIRST PHASE WILL BE ASKED TO SUBMIT A PHASE 2 TECHNICAL REQUEST FOR PROPOSAL (RFP). PHASE 2 WILL REQUIRE A FULL ON-SITE AUDIT OF TWO FACILITIES WITH APPROPRIATE DOCUMENTATION AND FINANCIAL CALCULATIONS. PHASE 2 EVALUATION PROCEDURES WILL NOT ONLY CONSIDER THE EXPERIENCE OF THE FIRM AND ITS ASSIGNED PERSONNEL, BUT WILL WEIGH THE TECHNICAL FEASIBILITY, ECONOMIC VIABILITY, SOUNDNESS OF PROJECT FINANCING ARRANGEMENTS, AND THE ESTIMATED

NET FINANCIAL BENEFIT TO OUR ORGANIZATION. THE _________[organization] INTENDS TO AWARD A NEGOTIATED CONTRACT TO ONE FIRM TO PROVIDE THE EQUIPMENT AND SERVICES UNDER TERMS AND CONDITIONS __________ [organization] CONSIDERS TO BE MOST FAVORABLE AMONG THOSE OFFERED.

Under III. L in the RFQ substitute instead of sample audit wording:

ENERGY AUDIT

THE SUCCESSFUL CONTRACTOR WILL BE REQUIRED TO DO DETAILED ENERGY AUDITS FOR EACH OF THE BUILDINGS INCLUDED IN ITS CONTRACT. THE PURPOSE OF THE RFP IS TO ASCERTAIN THE ESCO'S TECHNICAL EXPERTISE, AUDIT METHODOLOGY, ECONOMIC VIABILITY AND THE POTENTIAL NET FINANCIAL BENEFIT TO _________ [organization].

THE PROPOSAL WILL PROVIDE A BRIEF DESCRIPTION OF ANTICIPATED EXTENT OF SUCH AN AUDIT; INCLUDING THE LENGTH OF THE PAYBACK PERIOD, WHICH WILL BE AT LEAST THREE YEARS, THAT PROPOSER WILL USE AS A CEILING TO GOVERN MEASURES TO BE RECOMMENDED. ANY EXPECTED LIMITATIONS ON MEASURES TO BE CONSIDERED, NOT EXCLUDED BY LENGTH OF THE PAYBACK, MUST BE NOTED. ENERGY AUDIT DESCRIPTION SHOULD BE REFLECTED IN TEST AUDIT PROCEDURES AND THE AUDIT REPORT SHOULD BE REFERENCED AS APPROPRIATE.

TEST AUDIT

BIDDERS WILL BE REQUIRED TO SUBMIT A SAMPLE AUDIT PROCEDURE AND FINANCIAL CALCULATIONS. A COMPREHENSIVE ENERGY AUDIT WILL BE CONDUCTED AT _______ [name of facility] AT _______. [address] AND SUBMITTED, AS A SEPARATE DOCUMENT. THE FINDINGS, RECOMMENDATIONS AND CALCULATIONS FOR THIS SITE WILL REPRESENT A SAMPLE OF THE FIRM'S PROPOSED ENERGY AUDIT WORK. APPLICATION OF ALL PROPOSED SERVICES APPROPRIATE TO AN INDIVIDUAL FACILITY SHOULD BE DESCRIBED.

FINANCIAL CALCULATIONS MAY ASSUME WORK AT THE TEST SITE IS PART OF [X number of buildings, or all the facilities]; ALL OTHER PROCEDURES SHOULD BE SITE-SPECIFIC.

BIDDERS MAY SCHEDULE AN AUDIT OF THE TEST SITE AT THE PRE-PROPOSAL CONFERENCE, OR FOLLOWING THE CONFERENCE BY CALLING _______ AT (__)

SELECTION CRITERIA

[Add to RFQ criteria as appropriate. Suggested language is offered below.]

TECHNICAL APPROACH

COMPREHENSIVENESS OF PROPOSED AUDITS
— ANY LIMITATIONS ON MEASURE TO BE CONSIDERED
— LONGEST INDIVIDUAL PAYBACK
— LONGEST COMBINED PAYBACK
CLARITY OF METHODOLOGY AND SAMPLE CALCULATIONS

METHOD OF SELECTION OF ENERGY EFFICIENT EQUIPMENT

RANGE OF ENERGY CONSERVATION OPPORTUNITIES IDENTIFIED; NEW TECHNOLOGIES

INTERFACE OF RECOMMENDED EQUIPMENT TO EXISTING EQUIPMENT

ADEQUACY OF EQUIPMENT OR SERVICES TO ACCOMPLISH SCOPE OF WORK

QUALITY OF TEST SITE(S) REPORT

FINANCIAL CONSIDERATIONS

CALCULATIONS OF SAVINGS—PROCEDURES FOR ADDRESSING SUCH VARIABLES AS CHANGES IN BUILDING USE, UTILITY RATES, WEATHER FLUCTUATIONS

PROJECTED LEVEL OF TOTAL ENERGY SAVINGS

OTHER MODIFICATIONS

In converting the RFQ to an RFP, the document should be carefully reviewed for any cosmetic changes needed to make the solicitation consistent throughout. Other procedural changes, such as the time specified in the deadlines, should consider the longer time period required for ESCO to do a test audit.

Whatever process is used, the key to effective solicitation procedures is to ask for just the information that will enable the administration to effectively judge the proposers' qualifications and competence to meet the organization's identified needs as reflected in the criteria.

APPENDIX B

ENERGY SERVICE COMPANIES MARKETS AND FINANCING AVAILABLE

ENERGY SERVICE COMPANIES MARKETS AND FINANCING AVAILABLE

Companies (alpha order)	Markets Preferred									Financing Offered/Arranged*						
	Public Schools	Hosp.	Univ.	State Local Gov't.	Fed. Gov't.	Non-Profit	Commer-cial	Indus-trial	Other	Lease	Muni-Lease	Loan	Shared Savings	Chauf-fage	Guar. Lease	Other
CES/WAY	✓	✓	✓	✓	✓	✓	✓	✓	✓[a]	✓	✓	✓	✓		✓	
Central Hudson	✓	✓	✓	✓			✓	✓		✓	✓		✓		✓	✓[b]
Citizens Conservation					✓				✓[c]			✓	✓			
Co-Energy Group	✓		✓	✓	✓	✓	✓	✓		✓	✓	✓	✓		✓	
Demand-Side Resources	✓	✓	✓	✓	✓	✓	✓	✓	✓[d]				✓			
EAU Cogen	✓	✓	✓	✓	✓	✓	✓	✓	✓[e]	✓	✓	✓	✓	✓	✓	
Energy Investments	✓	✓	✓	✓	✓		✓	✓		✓		✓	✓			✓[f]
Energy Masters	✓	✓	✓	✓	✓	✓	✓	✓		✓	✓	✓	✓	✓	✓	
Enersave	✓	✓	✓				✓	✓		✓		✓	✓	✓	✓	
Financial Energy Management	✓	✓	✓	✓	✓		✓	✓		✓	✓	✓	✓		✓	
Freedom Energy Services	✓	✓	✓	✓	✓	✓	✓	✓	✓[g]	✓		✓	✓		✓	
HEC	✓	✓	✓	✓	✓	✓	✓	✓		✓	✓		✓		✓	✓[h]
Highland Energy Group	✓	✓	✓	✓	✓	✓	✓	✓	✓[i]							
Honeywell	✓	✓	✓	✓	✓	✓	✓	✓	✓[j]							✓[k]

Johnson Controls	✓	✓	✓	✓	✓	✓	✓	✓								
Kenetech Energy Management	✓	✓	✓	✓	✓	✓	✓	✓	✓l							
ML Systems		✓	✓	✓	✓		✓	✓	✓m							
Northeast Energy Service	✓	✓	✓	✓		✓	✓	✓								
S&C Thermal Systems	✓	✓	✓	✓	✓	✓	✓	✓	✓n							✓o
Sycom Enterprises	✓	✓	✓	✓	✓	✓	✓	✓								
Transphase Systems	✓	✓	✓	✓	✓		✓	✓								
Viron	✓	✓	✓	✓		✓	✓									
Vision Impact	✓	✓	✓	✓	✓	✓	✓	✓	✓p							

a) Housing authorities
b) Third-party financing
c) Multi-family, especially HUD-assisted, state-assisted, public housing
d) "All sectors"
e) Retail
f) "Variety of financing structures as needed and as prudent."
g) MIlitary
h) "Third-party grants, subsidies and loans."
i) Utility DSM Programs
j) Utilities
k) "Third party financing, service agreements with savings guarantee, cash sales agreements with savings guarantee."
l) Service industry
m) Retail
n) We serve all sectors
o) Third party financing
p) Retail

*Based on info provided in the NAESCO public bidders list, 1991.

ENERGY SERVICE COMPANIES[1]

COMPANY	SERVICES OFFERED
CES/Way International, Inc. 5308 Ashbrook Houston, TX 77081 Contact: Thomas K. Dreessen (713) 666-3541	Energy Audits Engineering Design Project Government Construction Ongoing Maintenance, Monitoring, Savings Guarantees
Central Hudson Enterprises Corp. 80 Washington Street Poughkeepsie, NY 12601 Contact: James Moss (914) 485-5770	Energy Audits Engineering Design Project Government Construction On-going Maintenance
Citizens Conservation Corporation 530 Atlantic Avenue Boston, MA 02210 Contact: Steve Morgan or Lillian Kamalay (617) 423-7900	Energy Audits Engineering Design Project Government Construction Performance Monitoring Resident Education Maintenance Staff Training
Co-Energy Group 725 Arizona Avenue, Suite #206 Santa Monica, CA 90401 Contact: Robert A. Freeman (213) 395-6767	Marketing Funding
Demand-Side Resources 111 West Washington, #1247 Chicago, IL 60602 Contact: Marc Vallen (312) 807-4848	Energy Audits Engineering Design Project Government Construction

Company	Services
EAU Cogenex Corporation Boott Mills South 100 Foot of John Street Lowell, MA 01852 Contact: Joseph S. Fitzpatrick (508) 441-0090	Energy Audits Engineering Design Project Government Construction Ongoing Maintenance
Energy Investments, Inc. 286 Congress Street Boston, MA 02210 Contact: Mary Myers Kauppila (617) 482-8228	Energy Audits Engineering Design Project Government Construction Ongoing Maintenance DSM Planning
Energy Masters Corporation 7701 College Blvd. Suite 240 Overland Park, KS 66210 Contact: Donald O. Smith (913) 469-5454	Energy Audits Engineering Design Project Government Construction Ongoing Maintenance Training Commissioning Follow-up Monitoring
Enersave, Inc. 164 Madison Avenue New York, NY 10016 Contact: Dennis T. Wilson (212) 684-6611	Energy Audits Engineering Design Project Government Construction Ongoing Maintenance
Financial Energy Management, Inc. 1625 Downing Street, Suite 101 Denver, CO 80218 Contact: James C. Crossman (303) 832-1920	Energy Audits Engineering Design Project Government Construction Ongoing Maintenance Electric Monitoring

Freedom Energy Services, Inc. 3368 Governor Drive, Suite 204F San Diego, CA 92122 Contact: Frank Coates (619) 756-5469	Energy Audits Engineering Design Construction Environmental Assessment
HEC, Inc. 24 Prime Parkway Natick, MA 01760 Contact: David S. Dayton (508) 653-0456	Energy Audits Engineering Design Project Government Construction Ongoing Maintenance
Highland Energy Group, Inc. 1536 Cole Blvd Golden, CO 80401 Contact: Leonard R. Rozek (303) 239-8455	Energy Audits Engineering Design Project Government Construction Ongoing Maintenance Building Simulation Software Development
Honeywell, Inc. MN27-4156 Honeywell Plaza Minneapolis, MN 55408 Contact: Kevin Kovak (612) 870-6557	Energy Audits Engineering Design Project Government Construction Ongoing Maintenance Indoor Air Quality Assessment Training Financing
Johnson Controls, Inc. 507 E. Michigan Street P.O. Box 423 Milwaukee, WI 53201 Contact: Bob Heller (414) 247-4557	Energy Audits Engineering Design Project Government Construction Ongoing Maintenance

Kenetech Energy Management, Inc.
15 New England Executive Park
Burlington, MA 01803

Contact: Cary C. Bullock
(617) 273-5894

Energy audits
Engineering Design
Project Government
Construction
Ongoing Maintenance
Consultants

ML Systems
465-2 Columbus Avenue
Valhalla, NY 10595

Contact: Don Bartley or
Phil Powers
(914) 741-0400

Energy Audits
Engineering Design
Construction
Ongoing Maintenance

Northeast Energy Services, Inc.
82 Florence Street
P.O. Box 496
Marlboro, MA 01752

Contact: George P. Sakellaris
(508) 485-3692

Energy Audits
Engineering Design
Project Government
Construction
Ongoing Maintenance
Finance

S & C Thermal Systems, Inc.
3314 Highway 162
P.O. Box 1246
Granite City, IL 62040

Contact: Roland Otte
(618) 452-3000

Energy Audits
Engineering Design
Project Government
Construction
Ongoing Maintenance
Operation/Chauffage
Concept

Sycom Enterprises
7475 Wisconsin Avenue
6th Floor
Bethesda, MD 20814

Contact: Richard D. Rathvon
(301) 718-6600

Energy Audits
Engineering Design
Project Government
Construction
Ongoing Maintenance

Transphase Systems, Inc. 15572 Computer Lane Huntington Beach, CA 92649 Contact: Doug Ames (714) 893-3920	Energy Audits Engineering Design Project Government Construction Ongoing Maintenance Manufacturing
Viron Corporation 216 NW Platte Valley Drive Riverside, MO 64150 Contact: George Diehl (816) 741-3500	Energy Audits Engineering Design Project Government Construction Ongoing Maintenance Training Follow-up Monitoring
Vision Impact Corporation 2970 Hartley Rd., Suite 204 Jacksonville, FL 32257 Contact: Russell W. Spitz (904) 262-6101	Energy Audits Engineering Design Project Government Construction Ongoing Maintenance

[1]Information based on NAESCO Bidder's List, 1991. There is no intent to suggest the above list is all inclusive.

INDEX